Advanced Topics of Topology

Edited by Francisco Bulnes

Published in London, United Kingdom

IntechOpen

Supporting open minds since 2005

Advanced Topics of Topology
http://dx.doi.org/10.5772/intechopen.94646
Edited by Francisco Bulnes

Contributors
Michael Baldado, Glaisa Tinguha Catalan, Roberto N. Padua, Punam Gupta, Sanjay Kumar Singh, P. Anbarasi Rodrigo, S. Maheswari, Lev Bukovský, Shuya Kanagawa, Kiyoyuki Tchizawa, Francisco Bulnes, Seyed Mohammad Ajdani, Yasuhiko Kamiyama

First published in London, United Kingdom, 2022 by IntechOpen
IntechOpen is the global imprint of INTECHOPEN LIMITED, registered in England and Wales, registration number: 11086078, 5 Princes Gate Court, London, SW7 2QJ, United Kingdom
Printed in Croatia

British Library Cataloguing-in-Publication Data
A catalogue record for this book is available from the British Library

Additional hard and PDF copies can be obtained from orders@intechopen.com

Advanced Topics of Topology
Edited by Francisco Bulnes
p. cm.
Print ISBN 978-1-80355-093-0
Online ISBN 978-1-80355-094-7
eBook (PDF) ISBN 978-1-80355-095-4

For EU product safety concerns:
IN TECH d.o.o., Prolaz Marije Krucifikse Kozulić 3, 51000 Rijeka, Croatia, info@intechopen.com or visit our website at intechopen.com.

Meet the editor

Dr. Francisco Bulnes, Ph.D., is IINAMEI Director, Mathematics Research Centre, Mexico. He is a member of various international committees of science and serves as a reviewer and editor for British and American journals of mathematics and physics. He is head of the Research Department, GI-TESCHA. He has published more than 100 papers and several books in mathematics and physics. Dr. Bulnes has many theories, theorems, and math objects to his credit. He has received various honors and awards from universities as well as governmental and non-governmental organizations. He received the Doctor Honoris Causa in Education Philosophy and is a Peace Ambassador for ODAEE in Frankfurt, Germany. He is also a distinguished member of the Czech Republic Mathematics Society (JCFM). He obtained two post-doctorates in Mathematics in Cuba and Russia. His research interests include electronics, microelectronics, and spintronics.

Contents

Preface

Topology, originally called "analysis situs," explores the properties and relations between geometric objects (or objects with certain set properties) in an ambient space. Topological invariants, which are properties that are preserved under a homeomorphism, are of great importance to the field of topology since many of them characterize geometric spaces and dynamical systems and define metrizability in spaces. Over six sections, this book discusses several types of topologies, including algebraic topology, differential topology, and symplectic topology, as well as topological spaces and manifolds, dimension theory, and the general study of topology. Modern topology developments focus on items and contents such as homotopy, cohomology, non-commutative rings, cobordisms, Lindelöf spaces, projective manifolds, connectivity, topos, and singular manifolds, all of which play a role in the development of deferent string theories and the topology of quantum field theories to the conceptual precision and understanding of the Universe. Also, in artificial intelligence and numerical simulation, topology is fundamental for defining spaces with adequate metrizability in dynamical systems and the design of artificial intelligence units of advanced automatons and the approach of androids to human behavior.

Dr. Francisco Bulnes, Ph.D, PostDocs, Doctor H. C., HonDSc, ZbMath, MathSci
Professor,
IINAMEI,
Research Department in Mathematics and Engineering,
TESCHA,
Chalco, Mexico

Section 1

Introduction

Chapter 1

Introductory Chapter: The Topology from Classic Studies until Its Last Frontiers

Francisco Bulnes

1. Introduction

The born of the topology is remounted with the perspective realized by Gottfried Leibniz, who in the seventeenth century envisioned the geometria situs and analysis situs [1]. After one of the most famous problems called Leonhard Euler's Seven Bridges of Königsberg problem and the polyhedron formula [2, 3] framework the formal initial study and the obtaining of the first results considering methods that after would give the graph theory [3], wherein a modern study associating algebra aspects is given the combinatorial topology, which establishes results-focused in the descomposibilities of vertices with certain invariance of co-dimensions on trees vertices as indirect graphs in which any two vertices are connected by exactly one path, thus using elements of commutative algebras with topology can be created results to its shell ability. Here also is more important the appearance of structures as topological groups to connect the vertex decomposability path complexes and the Dynkin graphs [4–6]. But the topology finds its greatest development when defined a topological space [7, 8] as a set endowed with a structure called topology which consists from a point of view purely orthodox of a system determined by metrics, norms, continuity elements, and defining continuous deformation of subspaces, where the deformations that are considered in topology are homeomorphisms [9, 10] and homotopies [11, 12]. Likewise, a property that is invariant under such deformations is a topological property. For example, the dimension concept, which endows a clear distinguishability between geometrical elements of different dimensions and the distinguishability between forms of two geometrical objects, can be made through compactness and connectedness. In a deep study arise the dimension theory [13, 14] as the topological theory of dimension of spaces.

However, the study of continuous applications and homeomorphisms legitimate the mappings between topological spaces and produces more specialized objects on the base of correspondences of these spaces, and under the resembling of Euclidean spaces near each point of a topological space called a manifold. Here, each point of an n-dimensional manifold has a neighborhood that is homeomorphic to the Euclidean space of dimension n.

IntechOpen

Introducing algebra to study topological spaces, we obtain algebraic invariants that classify topological spaces up to homeomorphism, though usually are classified in a complete way with the homotopy equivalence [15].

2. From coverings, cobordisms, homotopies, and topology shepeers until the theories of Stone-Cech compactification, and others, with the theory of rings of continuous functions, and more

A covering is a local homeomorphism. This relation establishes different topological properties depending on the object or topological mapping constructed for establishing lifting, deck (covering) transformation group, regular covering [16], G-covering [16–18] or even certain topological actions as the monodromy action [19], homomorphisms of fundamental groups, groupoids relations even Galois connection.

Likewise, a universal covering [15] $p : \mathrm{D} \rightarrow \mathrm{X}$, is regular, with the deck transformation group being isomorphic to the fundamental group $\pi_1(X)$.

The homotopy theory is a relative modern theory very useful in categories and schemes of categories, likewise as in the study of invariants as are homotopy groups, homology, and cohomology.

Today the most and biggest important problems in algebraic geometry and topology are focused on the study of higher categories, the p-adic groups theory, symplectic geometries, spectrum and generalized cohomology, motivic cohomology on tensor derived categories, among others. Even in the representation theory to classifying spaces of vector bundles, the homotopy operations result in determinant, where for a topological group G, the classifying space for principal G-bundles s a space BG, such that, for each topological space X, we have

$$[X, BG] = \{\text{principal } G - \text{bundle on } X\}/\sim, [f] \mapsto f^*EG, \qquad (1)$$

where here Brown's representability theorem guarantees the existence of classifying spaces.

The idea of classifying space that classifies principal bundles can be generalized and inducted. In the case of generalized cohomology, the classification is realized on the base contravariant functors from the category of spaces to the category of abelian groups that satisfies the axioms generalizing ordinary cohomology theory.

Likewise, for example in derived categories problems, the classifying space can be determined through a spectrum $\mathrm{Mod}_B\mathrm{CRing}_{\mathrm{A}//\mathrm{B}}$, which is defined by $X \mapsto \Omega_{X/\mathrm{A}} \otimes_X B$. In particular, the image of $\mathrm{A} \rightarrow B \rightarrow \mathrm{B}$, under this functor is $B \mapsto \Omega_{X/\mathrm{A}}$. As the important fact is necessary to consider that the derived tensor product is a regular tensor product [20]. Then the adjunction of categories $\mathrm{Ch}(B)_B \leftrightarrow s\mathrm{CRing}_{A//B}$, induces an adjunction to the level of homotopy categories [21, 22]. In this sense results very interesting the sphere homotopies used in generalized cohomology where the image of the functor is a sphere spectrum given, for example, by $S^0 \rightarrow S^1 \rightarrow S^2$. In this study arise important results of the homotopy theory as the homotopy excision theorem, the suspension theorem, etcetera. In the sphere, homotopy result interesting the study of invariants obtained correlating the stable homotopy theory and cobordisms [23, 24] obtained through these invariants between spheres and their compact groups, or a class of compact manifolds of the same dimension, where cobordism is an equivalence relation on those compact manifolds. A concrete application of it, is the compute of cohomologies or homologies considering oriented cobordism and complex cobordism. Likewise, in algebraic topology, the cobordism is a calculation method to determine cohomologies, for example, the homotopy groups [25–27] of spheres where spheres of various dimensions can wrap around each other, and finally result are related by homotopy invariants and topological invariants of the homotopy groups. Likewise, homotopy groups such as $\pi_{n+k}(S^n)$, result independent of the dimension sphere n, for $n \geq k + 2$. These are the stable homotopy groups of spheres that form a coefficient ring of a special theory of cohomology called stable cohomology theory. The unstable homotopy groups $n < \mathrm{k} + 2$ no form a ring, however, these have been tabulated to $n < 20$. Modern calculations using spectral sequences (method whose fundamental is homotopy) obtain more patterns to tabulation of homotopy groups to calculations in cohomology.

In a classic sense to the compactness [28] a Lindelöf space [28, 29] results in a suitable topological space due to that in a notion of compactness is required the existence of a finite sub-cover. Then in measure theory to some measure fields as the σ-field results Lindelöf. However, topologically what is beyond of the compactness for example, strong compactness[1], para-compactness, or pre-compactness. Even in many cases the compactness requires its generalization.

From specialized studies of compactification to establish universal mappings between topological spaces and compact Hausdorff spaces, arise fundamental concepts to functionality and the universal property. Likewise, a compact Hausdorff space can be characterized through homeomorphisms and their topological measure invariants can be used in integration theory both in integrals over spaces of infinite dimension, as integration in chains (giving place to the integral topology) and integral transforms.

However, in the compactificación problem arise specialized studies of intervals and sets in fuzzy theory and other kinds of sets to neutrosophic sets [30, 31]. Likewise, considering $\mathbb{P}$, a non-empty fixed set, we can say that a neutrosophic set Ħ on the universe $\mathbb{P}$ is defined as the space $\text{Ħ}\{\langle p, (t_{Ħ}(p), i_{Ħ}(p), f_{Ħ}(p))\rangle : p\epsilon P\}$ where $t_{Ħ}(p), i_{Ħ}(p), f_{Ħ}(p)$ represent the degree of membership function $t_{Ħ}(p)$, the degree of indeterminacy $i_{Ħ}(p)$ and the degree of non-membership function $f_{Ħ}(p)$ respectively for each element $p\epsilon P$ to the set Ħ. Also, $t_{Ħ}, i_{Ħ}, f_{Ħ}: \mathbb{P} \rightarrow {}^{-0,1^+}$ and $-0 \leq t_{Ħ}(p) + i_{Ħ}(p) + f_{Ħ}(p) \leq 3^+$. The set of all Neutrosophic sets over $\mathbb{P}$ is denoted by $\mathrm{N_{eu}}(\mathbb{P})$. Then is a generalization of the classic set, fuzzy set, interval-valued fuzzy set, intuitionistic fuzzy set, and paraconsistent set.

After, the separability is classified due to the separation axioms in topological spaces, having several classes of T_i-separability with $i = 0, 1, 2, 21/2, 3, 31/2, 4, 5, 6$[2]. Then different topological spaces are generated on the base of the separation axioms [32], and some other property related with every pair of topologically distinguishable points [28], for example, an accessible space or a space with Fréchet topology. Likewise, in any topological space we have, as properties of any two distinguishable points, the implications:

$$\text{Separated} \Rightarrow \text{topological distinguishable} \Rightarrow \text{distinct topological space}, \qquad (2)$$

Interesting embedding images of spaces arise called embedding separable metric spaces [28], for example, a separable metric space is homeomorphic to a subset of the Hilbert cube. In contraposition, also are of importance the non-separable spaces [29], where the elements of the space can be of bounded variation, or equipped with

---

[1] Noetherian topological space.

[2] T_0 (Kolmogorov); T_1 (Fréchet); T_2 (Hausdorff); $T_{2½}$ (Urysohn) completely; T_2 (completely Hausdorff); T_3 (regular Hausdorff); $T_{3½}$ (Tychonoff); T_4 (normal Hausdorff); T_5 (completely normal Hausdorff); T_6 (perfectly normal Hausdorff).

uncountable ordinals or even the spaces L^{∞}.Exist more non-separable spaces which relate metric spaces of density equal to infinite cardinals α, which are isometric to subspaces of continuous functions as $C([0,1]^{\alpha}, R)$.

3. Other specialized topology studies

Inside the differential topology, in the differentiable manifold context are developed many topological extensions of concepts as embedding, immersion, submersion, transversality, inclusion, epimorphism, diffeomorphisms, etc.

Likewise, the geometric properties of Clairaut anti-invariant submersions and Clairaut semi-invariant submersions [33, 34] whose total space is a Kähler, nearly Kähler manifold, are studied through certain conditions that make totally geodesic mappings. This is important in the aspects of connections and umbilical points of the Kähler manifolds [33], likewise, also the investigated conditions for the semi-invariant submersion to be a Clairaut mapping.

Also, the geometric topology is fundamental to study in the ambit of the high-dimensional topology [35], the characteristic classes [36] as the basic invariant, and surgery theory, which conform a key theory, in aspects more subtly, accurate, and specialized of the general topology.

Some generalizations have been mentioned considering beyond of “set of points” which could not be available. Then we consider the concept of the lattice of open sets as the basic notion of the theory [37]. This notion is an alternative, for example of the Grothendieck topologies [38] which are structures defined on arbitrary categories that allow the definition of sheaves (germ spaces) on those categories, and such that give the definition of general cohomology theories [39] or even specialized themes of cohomology as motivic cohomology of tensor cohomologies.

The applications are diverse, though in computer science and physics find more application in the topological data analysis [40, 41] and topological quantum field theory [42, 43] where knot theory [41], the theory of four-manifolds in algebraic topology [44], and to the theory of moduli spaces [45] regain importance.

Author details

Francisco Bulnes
IINAMEI, Research Department in Mathematics and Engineering, TESCHA, Chalco, Mexico

*Address all correspondence to: francisco.bulnes@tesch.edu.mx

References

[1] Poincaré H. Analysis situs. Journal de l'École Polytechnique. 1895;**1**(2):1-123

[2] Aigner M, Ziegler GM. Three Applications of Euler's Formula. Ch. 10 in Proofs from the Book. Berlin: Springer-Verlag; 1998

[3] Coxeter HSM. Euler's Formula and "Poincaré's Proof of Euler's Formula." §1.6 and Ch. 9 in Regular Polytopes. 3rd ed. New York: Dover; 1973. pp. 9-11 and 165-172

[4] Bondi A, Murray USR. "Graph Theory." North-Holland. N. Y., USA; 1976

[5] Diestel R. Graph Theory. Graduate Texts in Mathematics Book Series. N. Y., USA: Springer; 2017

[6] Humphreys JE. 48. Fundamental domain § Affine reflection groups. In: Reflection Groups and Coxeter Groups. United Kingdom: Cambridge University Press; 1990

[7] Berge C. Topological Spaces Including a Treatment of Multi-Valued Functions, Vector Spaces and Convexity. New York: Dover; 1997

[8] Hausdorff F. Grundzüge der Mengenlehre. Leipzig, Germany: von Veit; 1914

[9] Dijkstra Jan J. On homeomorphism groups and the compact-open topology. The American Mathematical Monthly. 2005;**112**(10):910. DOI: 10.2307/30037630

[10] Willard S. General Topology [1970]. 1st ed. Mineola, N.Y: Dover Publications; 2004

[11] Aubry M. Homotopy Theory and Models. Birkhäuser: Boston, MA; 1995

[12] Collins GP. The shapes of space. Scientific American. 2004;**291**:94-103

[13] Eisenbud D. Commutative algebra. In: With a view toward algebraic geometry, Graduate Texts in Mathematics. Vol. 150. New York: Springer-Verlag; 1995

[14] Weibel CA. An Introduction to Homological Algebra. Cambridge, United Kingdom: Cambridge University Press; 1995

[15] Hatcher A. Algebraic Topology. Cambridge: Cambridge University Press; 2002

[16] Brazas J. Semicoverings: A generalization of covering space theory. Homology, Homotopy and Applications. 2012;**14**(1):33-63. arXiv:1108.3021

[17] Taylor RL. Covering groups of nonconnected topological groups. Proceedings of the American Mathematical Society. 1954;**5**:753-768. DOI: 10.1090/S0002-9939-1954-0087028-0

[18] Brown R, Mucuk O. Covering groups of nonconnected topological groups revisited. Mathematical Proceedings of the Cambridge Philosophical Society. 1994;**115**(1): 97-110. arXiv:math/0009021

[19] König W, Sprekels J. Karl Weierstraß (1815–1897): Aspekte seines Lebens und Werkes. Berlin, Germany: Springer-Verlag; 2015. pp. 200-201

[20] Mazza C, Voevodsky V, Weibel C. Lecture Notes on Motivic Cohomology. Vol. 2. Cambridge MA, USA: AMS, Clay Mathematics Institute; 2006

[21] Beke T. Sheafifiable homotopy model categories. Mathematical Proceedings of the Cambridge Philosophical Society. 2000;**129**(3): 447-473. arXiv:math/0102087

[22] Dwyer WG, Spaliński J. Homotopy theories and model categories. In:

Handbook of Algebraic Topology. Amsterdam: North-Holland; 1995. pp. 73-126

[23] Michael FA. Bordism and cobordism. Mathematical Proceedings of the Cambridge Philosophical Society. 1961;**57**:200-208

[24] Milnor J. A survey of cobordism theory. L'Enseignement Mathématique. 1962;**8**:16-23

[25] Brown R. Groupoids and crossed objects in algebraic topology. Homology, Homotopy and Applications. 1999;**1**:1-78

[26] Čech E. Höherdimensionale Homotopiegruppen. Zürich: Verhandlungen des Internationalen Mathematikerkongress; 1932

[27] Kamps KH, Porter T. Abstract homotopy and simple homotopy theory. River Edge, NJ: World Scientific Publishing; 1997. DOI:10.1142/97898128 31989

[28] Willard S. General Topology. New York, USA: Dover Publications; 1970. ISBN 0-486-43479-6

[29] Engelking R. General Topology. Berlin: Heldermann Verlag; 1989

[30] Blessie Rebecca S, Francina Shalini A. Neutrosophic generalized regular contra continuity in neutrosophic topological spaces. Research in Advent Technology. 2019; **7**(2):761-765. E-ISSN:2321-9637

[31] Dhavaseelan R, Jafari S. Generalized neutrosophic closed sets. New trends in Neutrosophic Theory And Applications. 2017;**2**:261-273

[32] Willard S. General topology. Reading, Mass: Addison-Wesley Pub. Co.; 1970. ISBN 0-486-43479-6

[33] Nomizu K. Foundations of Differential Geometry, Interscience Tracts in Pure and Applied Mathematics. Vol. II. N.Y., USA: Interscience Publishers, Wiley & Sons; 1969

[34] Sahin B. Semi-invariant submersion from almost Hermitian manifolds. Canadian Mathematical Bulletin. 2011; **56**:173-182

[35] Sher RB, Daverman RJ. Handbook of Geometric Topology. North-Holland; 2002

[36] Chern S-S. Complex Manifolds Without Potential Theory. New Jersey, USA: Springer-Verlag Press; 1995

[37] Birkhoff G. Lattice Theory. Vol. 25. New York, NY: AMS Colloquium Publications; 1940. ISBN 978-0821810255

[38] Artin M. Grothendieck Topologies. Cambridge, MA: Harvard University, Dept. of Mathematics; 1962

[39] Switzer R. Algebraic Topology— Homology and Homotopy. New York, USA: Springer-Verlag; 1975. ISBN 3-540-42750-3

[40] Zadeh LA. "Fuzzy sets," In: Information and control. Vol. 8. Cambridge, United Kingdom: Cambridge University Press; 1965. pp. 338–353

[41] Deo N. Graph Theory with Applications to Engineering and Computer Science. New Jersey, USA: Prentice-Hall Inc; 1974

[42] Gukov S, Kapustin A. Topological quantum field theory, nonlocal operators, and gapped phases of gauge theories. Scimago. Journal of High Energy Physics Theory. (United Kingdom: Scimago). 2013;arXiv: 1307.4793

[43] Witten E. Topological quantum field theory. Communications in Mathematical Physics. 1988;**117**(3):353-386

[44] Maunder CRF. Algebraic Topology. London: Van Nostrand Reinhold; 1970. ISBN 0-486-69131-4

[45] Harris J, Morrison I. Moduli of curves. In: Graduate Texts in Mathematics. Vol. 187. New York: Springer Verlag; 1998

Section 2

Special Topological Sets and Their Continous Applications

Chapter 2

More Functions Associated with Neutrosophic gsα*- Closed Sets in Neutrosophic Topological Spaces

P. Anbarasi Rodrigo and S. Maheswari

Abstract

The concept of neutrosophic continuous function was very first introduced by A.A. Salama et al. The main aim of this paper is to introduce a new concept of Neutrosophic continuous function namely Strongly Neutrosophic gsα* - continuous functions, Perfectly Neutrosophic gsα* - continuous functions and Totally Neutrosophic gsα* - continuous functions in Neutrosophic topological spaces. These concepts are derived from strongly generalized neutrosophic continuous function and perfectly generalized neutrosophic continuous function. Several interesting properties and characterizations are derived and compared with already existing neutrosophic functions.

Keywords: Neutrosophic $gs\alpha^*$- closed set, Neutrosophic $gs\alpha^*$- open set, Strongly Neutrosophic $gs\alpha^*$- continuous function, Perfectly Neutrosophic $gs\alpha^*$- continuous function, Totally Neutrosophic $gs\alpha^*$- continuous function

1. Introduction

The concept of Neutrosophic set theory was introduced by F. Smarandache [1] and it comes from two concept, one is intuitionistic fuzzy sets introduced by K. Atanassov's [2] and the other is fuzzy sets introduced by L.A. Zadeh's [3]. It includes three components, truth, indeterminancy and false membership function. R. Dhavaseelan and S. Jafari [4] has discussed about the concept of strongly generalized neutrosophic continuous function. Further he also introduced the topic of perfectly generalized neutrosophic continuous function. The real life application of neutrosophic topology is applied in Information Systems, Applied Mathematics etc.

In this paper, we introduce some new concepts related to Neutrosophic $gs\alpha^* -$ continuous function namely Strongly Neutrosophic $gs\alpha^* -$ continuous function, Perfectly Neutrosophic $gs\alpha^* -$ continuous function, Totally Neutrosophic $gs\alpha^* -$ continuous function.

2. Preliminaries

Definition 2.1: [5] Let $\mathbb{P}$ be a non-empty fixed set. A Neutrosophic set H on the universe $\mathbb{P}$ is defined as $\mathrm{H} = \{\langle \boldsymbol{p}, (t_{\mathrm{H}}(\boldsymbol{p}), i_{\mathrm{H}}(\boldsymbol{p}), f_{\mathrm{H}}(\boldsymbol{p}))\rangle : \boldsymbol{p} \in \mathbb{P}\}$ where $t_{\mathrm{H}}(\boldsymbol{p}), i_{\mathrm{H}}(\boldsymbol{p}), f_{\mathrm{H}}(\boldsymbol{p})$ represent the degree of membership function $t_{\mathrm{H}}(\boldsymbol{p})$, the degree of indeterminacy $i_{\mathrm{H}}(\boldsymbol{p})$ and the degree of non-membership function $f_{\mathrm{H}}(\boldsymbol{p})$ respectively for each element $\boldsymbol{p} \in \mathbb{P}$ to the set H. Also, $t_{\mathrm{H}}, i_{\mathrm{H}}, f_{\mathrm{H}} : \mathbb{P} \to]^{-}0, 1^{+}[$ and $^{-}0$

$\leq t_{H}(p) + i_{H}(p) + f_{H}(p) \leq 3^{+}$. Set of all Neutrosophic set over $\mathbb{P}$ is denoted by $N_{eu}(\mathbb{P})$.

Definition 2.2: [8] Let $\mathbb{P}$ be a non-empty set. $A = \{\langle p, (t_A(p), i_A(p), f_A(p)) \rangle : p \in \mathbb{P}\}$ and $B = \{\langle p, (t_B(p), i_B(p), f_B(p)) \rangle : p \in \mathbb{P}\}$ are neutrosophic sets, then

i. $A \subseteq B$ if $t_A(p) \leq t_B(p), i_A(p) \leq i_B(p), f_A(p) \geq f_B(p)$ for all $p \in \mathbb{P}$.

ii. Intersection of two neutrosophic set A and B is defined as $A \cap B = \{\langle p, (\min(t_A(p), t_B(p)), \min(i_A(p), i_B(p)), \max(f_A(p), f_B(p))) \rangle : p \in \mathbb{P}\}$.

iii. Union of two neutrosophic set A and B is defined as $A \cup B = \{\langle p, (\max(t_A(p), t_B(p)), \max(i_A(p), i_B(p)), \min(f_A(p), f_B(p))) \rangle : p \in \mathbb{P}\}$.

iv. $A^c = \{\langle p, (f_A(p), 1 - i_A(p), t_A(p)) \rangle : p \in \mathbb{P}\}$.

v. $0_{N_{eu}} = \{\langle p, (0,0,1) \rangle : p \in \mathbb{P}\}$ and $1_{N_{eu}} = \{\langle p, (1,1,0) \rangle : p \in \mathbb{P}\}$.

Definition 2.3: [5] A neutrosophic topology ($N_{eu}T$) on a non-empty set $\mathbb{P}$ is a family $\tau_{N_{eu}}$ of neutrosophic sets in $\mathbb{P}$ satisfying the following axioms,

i. $0_{N_{eu}}, 1_{N_{eu}} \in \tau_{N_{eu}}$.

ii. $A_1 \cap A_2 \in \tau_{N_{eu}}$ for any $A_1, A_2 \in \tau_{N_{eu}}$.

iii. $\bigcup A_i \in \tau_{N_{eu}}$ for every family $\{A_i / i \in \Omega\} \subseteq \tau_{N_{eu}}$.

In this case, the ordered pair $(\mathbb{P}, \tau_{N_{eu}})$ or simply $\mathbb{P}$ is called a neutrosophic topological space ($N_{eu}TS$). The elements of $\tau_{N_{eu}}$ is neutrosophic open set ($N_{eu} - OS$) and $\tau_{N_{eu}}{}^c$ is neutrosophic closed set ($N_{eu} - CS$).

Definition 2.4: [6] A neutrosophic set A in a $N_{eu}TS$ $(\mathbb{P}, \tau_{N_{eu}})$ is called a neutrosophic generalized semi alpha star closed set ($N_{eu}gs\alpha^* - CS$) if $N_{eu}\alpha - int(N_{eu}\alpha - cl(A)) \subseteq N_{eu} - int(\mathcal{G})$, whenever $A \subseteq \mathcal{G}$ and $\mathcal{G}$ is $N_{eu}\alpha^* -$ open set.

Definition 2.5: [7] A neutrosophic topological space $(\mathbb{P}, \tau_{N_{eu}})$ is called a $N_{eu}gs\alpha^* - T_{1/2}$ space if every $N_{eu}gs\alpha^* - CS$ in $(\mathbb{P}, \tau_{N_{eu}})$ is a $N_{eu} - CS$ in $(\mathbb{P}, \tau_{N_{eu}})$.

Definition 2.6: A neutrosophic function $f : (\mathbb{P}, \tau_{N_{eu}}) \to (\mathbb{Q}, \sigma_{N_{eu}})$ is said to be

1. neutrosophic continuous [8] if the inverse image of each $N_{eu} - CS$ in $(\mathbb{Q}, \sigma_{N_{eu}})$ is a $N_{eu} - CS$ in $(\mathbb{P}, \tau_{N_{eu}})$.

2. $N_{eu}gs\alpha^* -$ continuous [7] if the inverse image of each neutrosophic closed set in $(\mathbb{Q}, \sigma_{N_{eu}})$ is a $N_{eu}gs\alpha^* -$ closed set in $(\mathbb{P}, \tau_{N_{eu}})$.

3. $N_{eu}gs\alpha^* -$ irresolute map [7] if the inverse image of each $N_{eu}gs\alpha^* -$ closed set in $(\mathbb{Q}, \sigma_{N_{eu}})$ is a $N_{eu}gs\alpha^* -$closed set in $(\mathbb{P}, \tau_{N_{eu}})$.

4. strongly neutrosophic continuous [4] if the inverse image of each neutrosophic set in $(\mathbb{Q}, \sigma_{N_{eu}})$ is both $N_{eu} - OS$ and $N_{eu} - CS$ in $(\mathbb{P}, \tau_{N_{eu}})$.

5. perfectly neutrosophic continuous [4] if the inverse image of each $N_{eu} - CS$ in $(\mathbb{Q}, \sigma_{N_{eu}})$ is both $N_{eu} - OS$ and $N_{eu} - CS$ in $(\mathbb{P}, \tau_{N_{eu}})$.

Definition 2.7: [9] Let $\tau_{N_{eu}} = \{0_{N_{eu}}, 1_{N_{eu}}\}$ is a neutrosophic topological space over $\mathbb{P}$. Then $(\mathbb{P}, \tau_{N_{eu}})$ is called neutrosophic discrete topological space.

Definition 2.8: A neutrosophic topological space $(\mathbb{P}, \tau_{N_{eu}})$ is called a neutrosophic clopen set ($N_{eu}-$clopen set) if it is both $N_{eu} - OS$ and $N_{eu} - CS$ in $(\mathbb{P}, \tau_{N_{eu}})$.

3. Strongly neutrosophic $gs\alpha^*$-continuous function

Definition 3.1: A neutrosophic function $f : (\mathbb{P}, \tau_{N_{eu}}) \to (\mathbb{Q}, \sigma_{N_{eu}})$ is said to be strongly $N_{eu}gs\alpha^* -$ continuous if the inverse image of every $N_{eu}gs\alpha^* - CS$ in $(\mathbb{Q}, \sigma_{N_{eu}})$ is a $N_{eu} - CS$ in $(\mathbb{P}, \tau_{N_{eu}})$. (ie) $f^{-1}(\mathbb{A})$ is a $N_{eu} - CS$ in $(\mathbb{P}, \tau_{N_{eu}})$ for every $N_{eu}gs\alpha^* - CS$ $\mathbb{A}$ in $(\mathbb{Q}, \sigma_{N_{eu}})$.

Theorem 3.2: Every strongly $N_{eu}gs\alpha^* -$ continuous is neutrosophic continuous, but not conversely.

Proof:

Let $f : (\mathbb{P}, \tau_{N_{eu}}) \to (\mathbb{Q}, \sigma_{N_{eu}})$ be any neutrosophic function. Let $\mathbb{A}$ be any $N_{eu} - CS$ in $(\mathbb{Q}, \sigma_{N_{eu}})$. Since every $N_{eu} - CS$ is $N_{eu}gs\alpha^* - CS$, then $\mathbb{A}$ is $N_{eu}gs\alpha^* - CS$ in $(\mathbb{Q}, \sigma_{N_{eu}})$. Since f is strongly $N_{eu}gs\alpha^* -$ continuous, then $f^{-1}(\mathbb{A})$ is $N_{eu} - CS$ in $(\mathbb{P}, \tau_{N_{eu}})$. Therefore, f is neutrosophic continuous.

Example 3.3: Let $\mathbb{P} = \{p\}$ and $\mathbb{Q} = \{q\}$. $\tau_{N_{eu}} = \{0_{N_{eu}}, 1_{N_{eu}}, \mathbb{A}\}$ and $\sigma_{N_{eu}} = \{0_{N_{eu}}, 1_{N_{eu}}, \mathbb{B}\}$ are $N_{eu}TS$ on $(\mathbb{P}, \tau_{N_{eu}})$ and $(\mathbb{Q}, \sigma_{N_{eu}})$ respectively. Also $\mathbb{A} = \{\langle p, (0.6, 0.4, 0.4)\rangle\}$ and $\mathbb{B} = \{\langle q, (0.4, 0.6, 0.2)\rangle\}$ are $N_{eu}(\mathbb{P})$ and $N_{eu}(\mathbb{Q})$. Define a map $f : (\mathbb{P}, \tau_{N_{eu}}) \to (\mathbb{Q}, \sigma_{N_{eu}})$ by $f(p) = q + 0.2$. Let $\mathbb{B}^c = \{\langle q, (0.2, 0.4, 0.4)\rangle\}$ be a $N_{eu} - CS$ in $(\mathbb{Q}, \sigma_{N_{eu}})$. Then $f^{-1}(\mathbb{B}^c) = \{\langle p, (0.4, 0.6, 0.6)\rangle\}$. Now, $N_{eu} - cl(f^{-1}(\mathbb{B}^c)) = \mathbb{A}^c \cap 1_{N_{eu}} = \mathbb{A}^c = f^{-1}(\mathbb{B}^c) \Rightarrow f^{-1}(\mathbb{B}^c)$ is $N_{eu} - CS$ in $(\mathbb{P}, \tau_{N_{eu}})$. Therefore, f is neutrosophic continuous, but f is not strongly $N_{eu}gs\alpha^* -$ continuous. Let $\mathbb{C} = \{\langle q, (0.1, 0.2, 0.8)\rangle\}$ be a $N_{eu}gs\alpha^* - CS$ in $(\mathbb{Q}, \sigma_{N_{eu}})$. Then $f^{-1}(\mathbb{C}) = \{\langle p, (0.3, 0.4, 1)\rangle\}$. Now $N_{eu} - cl(f^{-1}(\mathbb{C})) = \mathbb{A}^c \cap 1_{N_{eu}} = \mathbb{A}^c \neq f^{-1}(\mathbb{C}) \Rightarrow f^{-1}(\mathbb{C})$ is not $N_{eu} - CS$ in $(\mathbb{P}, \tau_{N_{eu}})$.

Theorem 3.4: Let $f : (\mathbb{P}, \tau_{N_{eu}}) \to (\mathbb{Q}, \sigma_{N_{eu}})$ be strongly $N_{eu}gs\alpha^* -$ continuous iff the inverse image of every $N_{eu}gs\alpha^* - OS$ in $(\mathbb{Q}, \sigma_{N_{eu}})$ is $N_{eu} - OS$ in $(\mathbb{P}, \tau_{N_{eu}})$.

Proof:

Assume that f is strongly $N_{eu}gs\alpha^* -$ continuous function. Let $\mathbb{A}$ be any $N_{eu}gs\alpha^* - OS$ in $(\mathbb{Q}, \sigma_{N_{eu}})$. Then $\mathbb{A}^c$ is $N_{eu}gs\alpha^* - CS$ in $(\mathbb{Q}, \sigma_{N_{eu}})$. Since f is strongly $N_{eu}gs\alpha^* -$ continuous, then $f^{-1}(\mathbb{A}^c)$ is $N_{eu} - CS$ in $(\mathbb{P}, \tau_{N_{eu}}) \Rightarrow (f^{-1}(\mathbb{A}))^c$ is $N_{eu} - CS$ in $(\mathbb{P}, \tau_{N_{eu}}) \Rightarrow f^{-1}(\mathbb{A})$ is $N_{eu} - OS$ in $(\mathbb{P}, \tau_{N_{eu}})$. Conversely, Let $\mathbb{A}$ be any $N_{eu}gs\alpha^* - CS$ in $(\mathbb{Q}, \sigma_{N_{eu}})$. Then $\mathbb{A}^c$ is $N_{eu}gs\alpha^* - OS$ in $(\mathbb{Q}, \sigma_{N_{eu}})$. By hypothesis, $f^{-1}(\mathbb{A}^c)$ is $N_{eu} - OS$ in $(\mathbb{P}, \tau_{N_{eu}}) \Rightarrow (f^{-1}(\mathbb{A}))^c$ is $N_{eu} - OS$ in $(\mathbb{P}, \tau_{N_{eu}}) \Rightarrow f^{-1}(\mathbb{A})$ is $N_{eu} - CS$ in $(\mathbb{P}, \tau_{N_{eu}})$. Therefore, f is strongly $N_{eu}gs\alpha^* -$ continuous.

Theorem 3.5: Every strongly $N_{eu}gs\alpha^* -$ continuous is $N_{eu}gs\alpha^* -$ continuous, but not conversely.

Proof:

Let $f : (\mathbb{P}, \tau_{N_{eu}}) \to (\mathbb{Q}, \sigma_{N_{eu}})$ be any neutrosophic function. Let $\mathbb{A}$ be any $N_{eu} - CS$ in $(\mathbb{Q}, \sigma_{N_{eu}})$. Then $\mathbb{A}$ is $N_{eu}gs\alpha^* - CS$ in $(\mathbb{Q}, \sigma_{N_{eu}})$. Since f is strongly $N_{eu}gs\alpha^* -$ continuous, then $f^{-1}(\mathbb{A})$ is a $N_{eu} - CS$ in $(\mathbb{P}, \tau_{N_{eu}}) \Rightarrow f^{-1}(\mathbb{A})$ is $N_{eu}gs\alpha^* - CS$ in $(\mathbb{P}, \tau_{N_{eu}})$. Therefore, f is $N_{eu}gs\alpha^* -$ continuous.

Example 3.6: Let $\mathbb{P} = \{p\}$ and $\mathbb{Q} = \{q\}$. $\tau_{N_{eu}} = \{0_{N_{eu}}, 1_{N_{eu}}, \mathbb{A}\}$ and $\sigma_{N_{eu}} = \{0_{N_{eu}}, 1_{N_{eu}}, \mathbb{B}\}$ are $N_{eu}TS$ on $(\mathbb{P}, \tau_{N_{eu}})$ and $(\mathbb{Q}, \sigma_{N_{eu}})$ respectively. Also $\mathbb{A} = \{\langle p, (0.4, 0.5, 0.7)\rangle\}$ and $\mathbb{B} = \{\langle q, (0.6, 0.8, 0.4)\rangle\}$ are $N_{eu}(\mathbb{P})$ and $N_{eu}(\mathbb{Q})$.

Define a map $f : (\mathbb{P}, \tau_{N_{eu}}) \to (\mathbb{Q}, \sigma_{N_{eu}})$ by $f(p) = q$. Let $B^c = \{\langle q, (0.4, 0.2, 0.6)\rangle\}$ be a $N_{eu} - CS$ in $(\mathbb{Q}, \sigma_{N_{eu}})$. Then $f^{-1}(B^c) = \{\langle p, (0.4, 0.2, 0.6)\rangle\}$. $N_{eu}\alpha^* - OS = N_{eu}\alpha - OS = \{0_{N_{eu}}, 1_{N_{eu}}, A\}$ and $N_{eu}\alpha - CS = \{0_{N_{eu}}, 1_{N_{eu}}, A^c\}$.
$N_{eu}\alpha\text{–}cl\left(f^{-1}(B^c)\right) = A^c \cap 1_{N_{eu}} = A^c$. Now, $N_{eu}\alpha - int\left(N_{eu}\alpha - cl\left(f^{-1}(B^c)\right)\right) = A \subseteq N_{eu} - int(1_{N_{eu}}) = 1_{N_{eu}}$, whenever $f^{-1}(B^c) \subseteq 1_{N_{eu}} \Rightarrow f^{-1}(B^c)$ is $N_{eu}gs\alpha^* - CS$ in $(\mathbb{P}, \tau_{N_{eu}})$. Therefore, f is $N_{eu}gs\alpha^*$ – continuous. But f is not strongly $N_{eu}gs\alpha^*$ –continuous. Let $C = \{\langle q, (0.3, 0.1, 0.7)\rangle\}$ be a $N_{eu}gs\alpha^* - CS$ in $(\mathbb{Q}, \sigma_{N_{eu}})$. Then $f^{-1}(C) = \{\langle p, (0.3, 0.1, 0.7)\rangle\}$. Now $N_{eu} - cl\left(f^{-1}(C)\right) = A^c \cap 1_{N_{eu}} = A^c \neq f^{-1}(C) \Rightarrow f^{-1}(C)$ is not $N_{eu} - CS$ in $(\mathbb{P}, \tau_{N_{eu}})$.

Theorem 3.7: Every strongly neutrosophic continuous is strongly $N_{eu}gs\alpha^*$ – continuous, but not conversely.

Proof:

Let $f : (\mathbb{P}, \tau_{N_{eu}}) \to (\mathbb{Q}, \sigma_{N_{eu}})$ be any neutrosophic function. Let A be any $N_{eu}gs\alpha^* - CS$ in $(\mathbb{Q}, \sigma_{N_{eu}})$. Since f is strongly neutrosophic continuous, then $f^{-1}(A)$ is both $N_{eu} - OS$ and $N_{eu} - CS$ in $(\mathbb{P}, \tau_{N_{eu}}) \Rightarrow f^{-1}(A)$ is $N_{eu} - CS$ in $(\mathbb{P}, \tau_{N_{eu}})$. Hence, f is strongly $N_{eu}gs\alpha^*$ – continuous.

Example 3.8: Let $\mathbb{P} = \{p\}$ and $\mathbb{Q} = \{q\}$. $\tau_{N_{eu}} = \{0_{N_{eu}}, 1_{N_{eu}}, A, C\}$ and $\sigma_{N_{eu}} = \{0_{N_{eu}}, 1_{N_{eu}}, B\}$ are N_{eu}TS on $(\mathbb{P}, \tau_{N_{eu}})$ and $(\mathbb{Q}, \sigma_{N_{eu}})$ respectively. Also $A = \{\langle p, (0.4, 0.6, 0.2)\rangle\}$, $C = \{\langle p, ([0.4, 1], [0.6, 1], [0, 0.2])\rangle\}$ and $B = \{\langle q, (0.4, 0.6, 0.2)\rangle\}$ are $N_{eu}(\mathbb{P})$ and $N_{eu}(\mathbb{Q})$. Define a map $f : (\mathbb{P}, \tau_{N_{eu}}) \to (\mathbb{Q}, \sigma_{N_{eu}})$ by $f(p) = q$. Let $T = \{\langle q, ([0, 0.2], [0, 0.4], [0.4, 1])\rangle\}$ be a $N_{eu}gs\alpha^* - CS$ in $(\mathbb{Q}, \sigma_{N_{eu}})$. Then $f^{-1}(T) = \{\langle p, ([0, 0.2], [0, 0.4], [0.4, 1])\rangle\}$. Now $N_{eu} - cl\left(f^{-1}(T)\right) = A^c \cap C^c \cap 1_{N_{eu}} = C^c = f^{-1}(T)$. Therefore, f is strongly $N_{eu}gs\alpha^*$ – continuous. But f is not strongly neutrosophic continuous. Let $E = \{\langle q, (0.4, 0.6, 0.2)\rangle\}$ be a neutrosophic set in $(\mathbb{Q}, \sigma_{N_{eu}})$. Then $f^{-1}(E) = \{\langle p, (0.4, 0.6, 0.2)\rangle\}$. Now $N_{eu} - int\left(f^{-1}(E)\right) = 0_{N_{eu}} \cup A = A = f^{-1}(E) \Rightarrow f^{-1}(E)$ is $N_{eu} - OS$ in $(\mathbb{P}, \tau_{N_{eu}})$. Also $N_{eu} - cl\left(f^{-1}(E)\right) = 1_{N_{eu}} \neq f^{-1}(E) \Rightarrow f^{-1}(E)$ is not $N_{eu} - CS$ in $(\mathbb{P}, \tau_{N_{eu}})$. Therefore, $f^{-1}(E)$ is not both $N_{eu} - OS$ and $N_{eu} - CS$ in $(\mathbb{P}, \tau_{N_{eu}})$.

Remark 3.9: Every strongly neutrosophic continuous is $N_{eu}gs\alpha^*$ – continuous, but not conversely. (by Theorem 3.5 & 3.7).

Theorem 3.10: Let $f : (\mathbb{P}, \tau_{N_{eu}}) \to (\mathbb{Q}, \sigma_{N_{eu}})$ be neutrosophic function and $(\mathbb{Q}, \sigma_{N_{eu}})$ be $N_{eu}gs\alpha^* - T_{1/2}$ space. Then the following are equivalent.

1. f is strongly $N_{eu}gs\alpha^*$ – continuous.

2. f is neutrosophic continuous.

Proof:

1. $\Rightarrow$ (2), Proof follows from theorem 3.2.

2. $\Rightarrow$ (1), Let A be any $N_{eu}gs\alpha^* - CS$ in $(\mathbb{Q}, \sigma_{N_{eu}})$. Since $(\mathbb{Q}, \sigma_{N_{eu}})$ is $N_{eu}gs\alpha^* - T_{1/2}$ space, then A is $N_{eu} - CS$ in $(\mathbb{Q}, \sigma_{N_{eu}})$. Since f is neutrosophic continuous, then $f^{-1}(A)$ is $N_{eu} - CS$ in $(\mathbb{P}, \tau_{N_{eu}})$. Therefore, f is strongly $N_{eu}gs\alpha^*$ – continuous.

Theorem 3.11: Let $f : (\mathbb{P}, \tau_{N_{eu}}) \to (\mathbb{Q}, \sigma_{N_{eu}})$ be $N_{eu}gs\alpha^*$ – continuous. Both $(\mathbb{P}, \tau_{N_{eu}})$ and $(\mathbb{Q}, \sigma_{N_{eu}})$ are $N_{eu}gs\alpha^* - T_{1/2}$ space, then f is strongly $N_{eu}gs\alpha^*$ – continuous.

Proof:

Let A be any $N_{eu}gs\alpha^* - CS$ in $(\mathbb{Q}, \sigma_{N_{eu}})$. Since $(\mathbb{Q}, \sigma_{N_{eu}})$ is $N_{eu}gs\alpha^* - T_{1/2}$ space, then A is $N_{eu} - CS$ in $(\mathbb{Q}, \sigma_{N_{eu}})$. Since f is $N_{eu}gs\alpha^* -$ continuous, then $f^{-1}(A)$ is $N_{eu}gs\alpha^* - CS$ in $(\mathbb{P}, \tau_{N_{eu}})$. Since $(\mathbb{P}, \tau_{N_{eu}})$ is $N_{eu}gs\alpha^* - T_{1/2}$ space, then $f^{-1}(A)$ is $N_{eu} - CS$ in $(\mathbb{P}, \tau_{N_{eu}})$. Therefore, f is strongly $N_{eu}gs\alpha^* -$ continuous.

Theorem 3.12: Let $f : (\mathbb{P}, \tau_{N_{eu}}) \rightarrow (\mathbb{Q}, \sigma_{N_{eu}})$ be strongly $N_{eu}gs\alpha^* -$ continuous, then f is $N_{eu}gs\alpha^* -$ irresolute.

Proof:

Let A be any $N_{eu}gs\alpha^* - CS$ in $(\mathbb{Q}, \sigma_{N_{eu}})$. Since f is strongly $N_{eu}gs\alpha^* -$ continuous, then $f^{-1}(A)$ is $N_{eu} - CS$ in $(\mathbb{P}, \tau_{N_{eu}}) \Rightarrow f^{-1}(A)$ is $N_{eu}gs\alpha^* - CS$ in $(\mathbb{P}, \tau_{N_{eu}})$. Hence, f is $N_{eu}gs\alpha^* -$ irresolute.

Theorem 3.13: Let $f : (\mathbb{P}, \tau_{N_{eu}}) \rightarrow (\mathbb{Q}, \sigma_{N_{eu}})$ be $N_{eu}gs\alpha^* -$ irresolute and $(\mathbb{P}, \tau_{N_{eu}})$ be $N_{eu}gs\alpha^* - T_{1/2}$ space, then f is strongly $N_{eu}gs\alpha^* -$ continuous.

Proof:

Let A be any $N_{eu}gs\alpha^* - CS$ in $(\mathbb{Q}, \sigma_{N_{eu}})$. Since f is $N_{eu}gs\alpha^* -$ irresolute, then $f^{-1}(A)$ is $N_{eu}gs\alpha^* - CS$ in $(\mathbb{P}, \tau_{N_{eu}})$. Since $(\mathbb{P}, \tau_{N_{eu}})$ is $N_{eu}gs\alpha^* - T_{1/2}$ space, then $f^{-1}(A)$ is $N_{eu} - CS$ in $(\mathbb{P}, \tau_{N_{eu}})$. Therefore, f is strongly $N_{eu}gs\alpha^* -$ continuous.

Theorem 3.14: Let $f : (\mathbb{P}, \tau_{N_{eu}}) \rightarrow (\mathbb{Q}, \sigma_{N_{eu}})$ and $g : (\mathbb{Q}, \sigma_{N_{eu}}) \rightarrow (\mathbb{R}, \gamma_{N_{eu}})$ be strongly $N_{eu}gs\alpha^* -$ continuous, then $gof : (\mathbb{P}, \tau_{N_{eu}}) \rightarrow (\mathbb{R}, \gamma_{N_{eu}})$ is strongly $N_{eu}gs\alpha^* -$ continuous.

Proof:

Let A be any $N_{eu}gs\alpha^* - CS$ in $(\mathbb{R}, \gamma_{N_{eu}})$. Since g is strongly $N_{eu}gs\alpha^* -$ continuous, then $g^{-1}(A)$ is $N_{eu} - CS$ in $(\mathbb{Q}, \sigma_{N_{eu}}) \Rightarrow g^{-1}(A)$ is $N_{eu}gs\alpha^* - CS$ in $(\mathbb{Q}, \sigma_{N_{eu}})$. Since f is strongly $N_{eu}gs\alpha^* -$ continuous, then $f^{-1}(g^{-1}(A)) = (gof)^{-1}(A)$ is $N_{eu} - CS$ in $(\mathbb{P}, \tau_{N_{eu}})$. Therefore, gof is strongly $N_{eu}gs\alpha^* -$ continuous.

Theorem 3.15: Let $f : (\mathbb{P}, \tau_{N_{eu}}) \rightarrow (\mathbb{Q}, \sigma_{N_{eu}})$ be strongly $N_{eu}gs\alpha^* -$ continuous and $g : (\mathbb{Q}, \sigma_{N_{eu}}) \rightarrow (\mathbb{R}, \gamma_{N_{eu}})$ be $N_{eu}gs\alpha^* -$continuous, then $gof : (\mathbb{P}, \tau_{N_{eu}}) \rightarrow (\mathbb{R}, \gamma_{N_{eu}})$ is neutrosophic continuous.

Proof:

Let A be any $N_{eu} - CS$ in $(\mathbb{R}, \gamma_{N_{eu}})$. Since g is $N_{eu}gs\alpha^* -$ continuous, then $g^{-1}(A)$ is $N_{eu}gs\alpha^* - CS$ in $(\mathbb{Q}, \sigma_{N_{eu}})$. Since f is strongly $N_{eu}gs\alpha^* -$ continuous, then $f^{-1}(g^{-1}(A)) = (gof)^{-1}(A)$ is $N_{eu} - CS$ in $(\mathbb{P}, \tau_{N_{eu}})$. Therefore, gof is neutrosophic continuous.

Theorem 3.16: Let $f : (\mathbb{P}, \tau_{N_{eu}}) \rightarrow (\mathbb{Q}, \sigma_{N_{eu}})$ be strongly $N_{eu}gs\alpha^* -$ continuous and $g : (\mathbb{Q}, \sigma_{N_{eu}}) \rightarrow (\mathbb{R}, \gamma_{N_{eu}})$ be $N_{eu}gs\alpha^* -$ irresolute, then $gof : (\mathbb{P}, \tau_{N_{eu}}) \rightarrow (\mathbb{R}, \gamma_{N_{eu}})$ is strongly $N_{eu}gs\alpha^* -$ continuous.

Proof:

Let A be any $N_{eu}gs\alpha^* - CS$ in $(\mathbb{R}, \gamma_{N_{eu}})$. Since g is $N_{eu}gs\alpha^* -$ irresolute, then $g^{-1}(A)$ is $N_{eu}gs\alpha^* - CS$ in $(\mathbb{Q}, \sigma_{N_{eu}})$. Since f is strongly $N_{eu}gs\alpha^* -$ continuous, then $f^{-1}(g^{-1}(A)) = (gof)^{-1}(A)$ is $N_{eu} - CS$ in $(\mathbb{P}, \tau_{N_{eu}})$. Therefore, gof is strongly $N_{eu}gs\alpha^* -$ continuous.

Theorem 3.17: Let $f : (\mathbb{P}, \tau_{N_{eu}}) \rightarrow (\mathbb{Q}, \sigma_{N_{eu}})$ be $N_{eu}gs\alpha^* -$ continuous and g: $(\mathbb{Q}, \sigma_{N_{eu}}) \rightarrow (\mathbb{R}, \gamma_{N_{eu}})$ be strongly $N_{eu}gs\alpha^* -$continuous, then $gof : (\mathbb{P}, \tau_{N_{eu}}) \rightarrow (\mathbb{R}, \gamma_{N_{eu}})$ is $N_{eu}gs\alpha^* -$ irresolute.

Proof:

Let A be any $N_{eu}gs\alpha^* - CS$ in $(\mathbb{R}, \gamma_{N_{eu}})$. Since g is strongly $N_{eu}gs\alpha^* -$ continuous, then $g^{-1}(A)$ is $N_{eu} - CS$ in $(\mathbb{Q}, \sigma_{N_{eu}})$. Since f is $N_{eu}gs\alpha^* -$ continuous, then $f^{-1}(g^{-1}(A)) = (gof)^{-1}(A)$ is $N_{eu}gs\alpha^* - CS$ in $(\mathbb{P}, \tau_{N_{eu}})$. Hence, gof is $N_{eu}gs\alpha^* -$ irresolute.

Theorem 3.18: Let $f : (\mathbb{P}, \tau_{N_{eu}}) \to (\mathbb{Q}, \sigma_{N_{eu}})$ be neutrosophic continuous and $g : (\mathbb{Q}, \sigma_{N_{eu}}) \to (\mathbb{R}, \gamma_{N_{eu}})$ be strongly $N_{eu}gs\alpha^* -$ continuous, then $gof : (\mathbb{P}, \tau_{N_{eu}}) \to (\mathbb{R}, \gamma_{N_{eu}})$ is strongly $N_{eu}gs\alpha^* -$ continuous.

Proof:

Let A be any $N_{eu}gs\alpha^* - CS$ in $(\mathbb{R}, \gamma_{N_{eu}})$. Since g is strongly $N_{eu}gs\alpha^* -$ continuous, then $g^{-1}(A)$ is $N_{eu} - CS$ in $(\mathbb{Q}, \sigma_{N_{eu}})$. Since f is neutrosophic continuous, then $f^{-1}(g^{-1}(A)) = (gof)^{-1}(A)$ is $N_{eu} - CS$ in $(\mathbb{P}, \tau_{N_{eu}})$. Hence, gof is strongly $N_{eu}gs\alpha^* -$ continuous.

Inter-relationship 3.19:

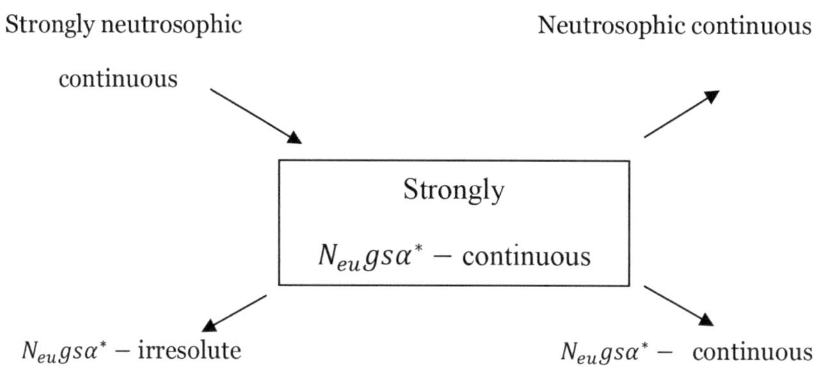

4. Perfectly neutrosophic $gs\alpha^*$-continuous function

Definition 4.1: A neutrosophic function $f : (\mathbb{P}, \tau_{N_{eu}}) \to (\mathbb{Q}, \sigma_{N_{eu}})$ is said to be perfectly $N_{eu}gs\alpha^* -$ continuous if the inverse image of every $N_{eu}gs\alpha^* - CS$ in $(\mathbb{Q}, \sigma_{N_{eu}})$ is both $N_{eu} - OS$ and $N_{eu} - CS$ (ie, $N_{eu} -$ clopen set) in $(\mathbb{P}, \tau_{N_{eu}})$.

Theorem 4.2: Every perfectly $N_{eu}gs\alpha^* -$ continuous is strongly $N_{eu}gs\alpha^* -$ continuous, but not conversely.

Proof:

Let $f : (\mathbb{P}, \tau_{N_{eu}}) \to (\mathbb{Q}, \sigma_{N_{eu}})$ be any neutrosophic function. Let A be any $N_{eu}gs\alpha^* - CS$ in $(\mathbb{Q}, \sigma_{N_{eu}})$. Since f is perfectly $N_{eu}gs\alpha^* -$ continuous, then $f^{-1}(A)$ is both $N_{eu} - OS$ and $N_{eu} - CS$ in $(\mathbb{P}, \tau_{N_{eu}}) \Rightarrow f^{-1}(A)$ is $N_{eu} - CS$ in $(\mathbb{P}, \tau_{N_{eu}})$. Therefore, f is strongly $N_{eu}gs\alpha^* -$ continuous.

Example 4.3: Let $\mathbb{P} = \{\boldsymbol{p}\}$ and $\mathbb{Q} = \{\boldsymbol{q}\}$. $\tau_{N_{eu}} = \{0_{N_{eu}}, 1_{N_{eu}}, A, C\}$ and $\sigma_{N_{eu}} = \{0_{N_{eu}}, 1_{N_{eu}}, B\}$ are N_{eu}TS on $(\mathbb{P}, \tau_{N_{eu}})$ and $(\mathbb{Q}, \sigma_{N_{eu}})$ respectively. Also $A = \{\langle \boldsymbol{p}, (0.7, 0.8, 0.3)\rangle\}$, $C = \{\langle \boldsymbol{p}, ([0.7, 1], [0.8, 1], [0, 0.3])\rangle\}$ and $B = \{\langle \boldsymbol{q}, (0.7, 0.8, 0.3)\rangle\}$ are $N_{eu}(\mathbb{P})$ and $N_{eu}(\mathbb{Q})$. Define a map $f : (\mathbb{P}, \tau_{N_{eu}}) \to (\mathbb{Q}, \sigma_{N_{eu}})$ by $f(\boldsymbol{p}) = \boldsymbol{q}$. Let $T = \{\langle \boldsymbol{q}, ([0, 0.3], [0, 0.2], [0.7, 1])\rangle\}$ be a $N_{eu}gs\alpha^* - CS$ in $(\mathbb{Q}, \sigma_{N_{eu}})$. Then $f^{-1}(T) = \{\langle \boldsymbol{p}, ([0, 0.3], [0, 0.2], [0.7, 1])\rangle\}$. Now $N_{eu} - cl(f^{-1}(T)) = A^c \cap C^c \cap 1_{N_{eu}} = C^c = f^{-1}(T)$. Therefore, f is strongly $N_{eu}gs\alpha^* -$ continuous. But f is not perfectly $N_{eu}gs\alpha^* -$ continuous, because $f^{-1}(T)$ is not both $N_{eu} - OS$ and $N_{eu} - CS$ in $(\mathbb{P}, \tau_{N_{eu}})$. Since, $N_{eu} - int(f^{-1}(T)) = 0_{N_{eu}} \neq f^{-1}(T) \Rightarrow f^{-1}(T)$ is not $N_{eu} - CS$ in $(\mathbb{P}, \tau_{N_{eu}})$. Therefore, $f^{-1}(T)$ is not both $N_{eu} - OS$ and $N_{eu} - CS$ in $(\mathbb{P}, \tau_{N_{eu}})$.

Theorem 4.4: Every perfectly $N_{eu}gs\alpha^* -$ continuous is perfectly neutrosophic continuous, but not conversely.

Proof:

Let $f : (\mathbb{P}, \tau_{N_{eu}}) \rightarrow (\mathbb{Q}, \sigma_{N_{eu}})$ be any neutrosophic function. Let A be any $N_{eu} - CS$ in $(\mathbb{Q}, \sigma_{N_{eu}})$. Then A is $N_{eu}gs\alpha^* - CS$ in $(\mathbb{Q}, \sigma_{N_{eu}})$. Since f is perfectly $N_{eu}gs\alpha^* -$ continuous, then $f^{-1}(A)$ is both $N_{eu} - OS$ and $N_{eu} - CS$ in $(\mathbb{P}, \tau_{N_{eu}})$. Therefore, f is perfectly neutrosophic continuous.

Example 4.5: Let $\mathbb{P} = \{p\}$ and $\mathbb{Q} = \{q\}$. $\tau_{N_{eu}} = \{0_{N_{eu}}, 1_{N_{eu}}, A, C, E\}$ and $\sigma_{N_{eu}} = \{0_{N_{eu}}, 1_{N_{eu}}, B\}$ are N_{eu}TS on $(\mathbb{P}, \tau_{N_{eu}})$ and $(\mathbb{Q}, \sigma_{N_{eu}})$ respectively. Also $A = \{\langle p, (0.4, 0.2, 0.6)\rangle\}$, $C = \{\langle p, (0.6, 0.8, 0.4)\rangle\}$, $E = \{\langle p, ([0, 0.4], [0, 0.2], [0.6, 1])\rangle\}$ and $B = \{\langle q, (0.6, 0.8, 0.4)\rangle\}$ are $N_{eu}(\mathbb{P})$ and $N_{eu}(\mathbb{Q})$. Define a map $f : (\mathbb{P}, \tau_{N_{eu}}) \rightarrow (\mathbb{Q}, \sigma_{N_{eu}})$ by $f(p) = q$. Let $B^c = \{\langle q, (0.4, 0.2, 0.6)\rangle\}$ be a $N_{eu} - CS$ in $(\mathbb{Q}, \sigma_{N_{eu}})$. Then $f^{-1}(B^c) = \{\langle p, (0.4, 0.2, 0.6)\rangle\}$. Now $N_{eu} - cl(f^{-1}(B^c)) = A^c \cap C^c \cap E^c \cap 1_{N_{eu}} = C^c = f^{-1}(B^c)$. Also, $N_{eu} - int(f^{-1}(B^c)) = A \cup E \cup 0_{N_{eu}} = A = f^{-1}(B^c) \Rightarrow f^{-1}(B^c)$ is both $N_{eu} - OS$ and $N_{eu} - CS$ in $(\mathbb{P}, \tau_{N_{eu}})$. Therefore, f is perfectly neutrosophic continuous. But f is not perfectly $N_{eu}gs\alpha^* -$ continuous. Let $T = \{\langle q, ([0, 0.4], [0, 0.2], [0.6, 1])\rangle\}$ be $N_{eu}gs\alpha^* - CS$ in $(\mathbb{Q}, \sigma_{N_{eu}})$. Then $f^{-1}(T) = \{\langle p, ([0, 0.4], [0, 0.2], [0.6, 1])\rangle\}$. Since, $N_{eu} - int(f^{-1}(T)) = E \cup 0_{N_{eu}} = E = f^{-1}(T)$ $\Rightarrow f^{-1}(T)$ is $N_{eu} - OS$ in $(\mathbb{P}, \tau_{N_{eu}})$. Also, $N_{eu} - cl(f^{-1}(T)) = A^c \cap C^c \cap E^c \cap 1_{N_{eu}} = C^c \neq f^{-1}(T) \Rightarrow f^{-1}(T)$ is not $N_{eu} - CS$ in $(\mathbb{P}, \tau_{N_{eu}})$. Therefore, $f^{-1}(T)$ is not both $N_{eu} - OS$ and $N_{eu} - CS$ in $(\mathbb{P}, \tau_{N_{eu}})$.

Theorem 4.6: Let $f : (\mathbb{P}, \tau_{N_{eu}}) \rightarrow (\mathbb{Q}, \sigma_{N_{eu}})$ be perfectly $N_{eu}gs\alpha^* -$ continuous iff the inverse image of every $N_{eu}gs\alpha^* - OS$ in $(\mathbb{Q}, \sigma_{N_{eu}})$ is both $N_{eu} - OS$ and $N_{eu} - CS$ in $(\mathbb{P}, \tau_{N_{eu}})$.

Proof:

Assume that f is perfectly $N_{eu}gs\alpha^* -$ continuous function. Let A be any $N_{eu}gs\alpha^* - OS$ in $(\mathbb{Q}, \sigma_{N_{eu}})$. Then A^c is $N_{eu}gs\alpha^* - CS$ in $(\mathbb{Q}, \sigma_{N_{eu}})$. Since f is perfectly $N_{eu}gs\alpha^* -$ continuous, then $f^{-1}(A^c)$ is both $N_{eu} - OS$ and $N_{eu} - CS$ in $(\mathbb{P}, \tau_{N_{eu}}) \Rightarrow (f^{-1}(A))^c$ is both $N_{eu} - OS$ and $N_{eu} - CS$ in $(\mathbb{P}, \tau_{N_{eu}}) \Rightarrow f^{-1}(A)$ is both $N_{eu} - OS$ and $N_{eu} - CS$ in $(\mathbb{P}, \tau_{N_{eu}})$. Conversely, Let A be any $N_{eu}gs\alpha^* - CS$ in $(\mathbb{Q}, \sigma_{N_{eu}})$. Then A^c is $N_{eu}gs\alpha^* - OS$ in $(\mathbb{Q}, \sigma_{N_{eu}})$. By hypothesis, $f^{-1}(A^c)$ is both $N_{eu} - OS$ and $N_{eu} - CS$ in $(\mathbb{P}, \tau_{N_{eu}}) \Rightarrow (f^{-1}(A))^c$ is both $N_{eu} - OS$ and $N_{eu} - CS$ in $(\mathbb{P}, \tau_{N_{eu}}) \Rightarrow f^{-1}(A)$ is both $N_{eu} - OS$ and $N_{eu} - CS$ in $(\mathbb{P}, \tau_{N_{eu}})$. Therefore, f is perfectly $N_{eu}gs\alpha^* -$ continuous.

Theorem 4.7: Let $(\mathbb{P}, \tau_{N_{eu}})$ be a neutrosophic discrete topological space and $(\mathbb{Q}, \sigma_{N_{eu}})$ be any neutrosophic topological space. Let $f : (\mathbb{P}, \tau_{N_{eu}}) \rightarrow (\mathbb{Q}, \sigma_{N_{eu}})$ be a neutrosophic function, then the following statements are true.

1. f is strongly $N_{eu}gs\alpha^* -$ continuous.

2. f is perfectly $N_{eu}gs\alpha^* -$ continuous.

Proof:

1. $\Rightarrow$ (2), Let A be any $N_{eu}gs\alpha^* - CS$ in $(\mathbb{Q}, \sigma_{N_{eu}})$. Since f is strongly $N_{eu}gs\alpha^* -$ continuous, then $f^{-1}(A)$ is $N_{eu} - CS$ in $(\mathbb{P}, \tau_{N_{eu}})$. Since $(\mathbb{P}, \tau_{N_{eu}})$ is neutrosophic discrete topological space, then $f^{-1}(A)$ is $N_{eu} - OS$ in $(\mathbb{P}, \tau_{N_{eu}}) \Rightarrow f^{-1}(A)$ is both $N_{eu} - OS$ and $N_{eu} - CS$ in $(\mathbb{P}, \tau_{N_{eu}})$. Therefore, f is perfectly $N_{eu}gs\alpha^* -$ continuous.

2. $\Rightarrow$ (1), Let A be any $N_{eu}gs\alpha^* - CS$ in $(\mathbb{Q}, \sigma_{N_{eu}})$. Since f is perfectly $N_{eu}gs\alpha^* -$ continuous, then $f^{-1}(A)$ is both $N_{eu} - OS$ and $N_{eu} - CS$ in $(\mathbb{P}, \tau_{N_{eu}}) \Rightarrow f^{-1}(A)$ is $N_{eu} - CS$ in $(\mathbb{P}, \tau_{N_{eu}})$. Therefore, f is strongly $N_{eu}gs\alpha^* -$ continuous.

Theorem 4.8: Let $f:(\mathbb{P},\tau_{N_{eu}})\to(\mathbb{Q},\sigma_{N_{eu}})$ be perfectly neutrosophic continuous and $(\mathbb{Q},\sigma_{N_{eu}})$ be $N_{eu}gs\alpha^* - T_{1/2}$ space, then f is perfectly $N_{eu}gs\alpha^*$ – continuous.

Proof:

Let A be any $N_{eu}gs\alpha^* - CS$ in $(\mathbb{Q},\sigma_{N_{eu}})$. Since $(\mathbb{Q},\sigma_{N_{eu}})$ is $N_{eu}gs\alpha^* - T_{1/2}$ space, then A is $N_{eu} - CS$ in $(\mathbb{Q},\sigma_{N_{eu}})$. Since f is perfectly neutrosophic continuous, then $f^{-1}(A)$ is both $N_{eu} - OS$ and $N_{eu} - CS$ in $(\mathbb{P},\tau_{N_{eu}})$. Therefore, f is perfectly $N_{eu}gs\alpha^*$ – continuous.

Theorem 4.9: Let $f:(\mathbb{P},\tau_{N_{eu}})\to(\mathbb{Q},\sigma_{N_{eu}})$ and $g:(\mathbb{Q},\sigma_{N_{eu}})\to(\mathbb{R},\gamma_{N_{eu}})$ be perfectly $N_{eu}gs\alpha^*$ – continuous, then $gof:(\mathbb{P},\tau_{N_{eu}})\to(\mathbb{R},\gamma_{N_{eu}})$ is perfectly $N_{eu}gs\alpha^*$ – continuous.

Proof:

Let A be any $N_{eu}gs\alpha^* - CS$ in $(\mathbb{R},\gamma_{N_{eu}})$. Since g is perfectly $N_{eu}gs\alpha^*$ – continuous, then $g^{-1}(A)$ is both $N_{eu} - OS$ and $N_{eu} - CS$ in $(\mathbb{Q},\sigma_{N_{eu}}) \Rightarrow g^{-1}(A)$ is $N_{eu}gs\alpha^* - CS$ in $(\mathbb{Q},\sigma_{N_{eu}})$. Since f is perfectly $N_{eu}gs\alpha^*$ – continuous, then $f^{-1}(g^{-1}(A)) = (gof)^{-1}(A)$ is both $N_{eu} - OS$ and $N_{eu} - CS$ in $(\mathbb{P},\tau_{N_{eu}})$. Therefore, gof is perfectly $N_{eu}gs\alpha^*$ – continuous.

Theorem 4.10: Let $f:(\mathbb{P},\tau_{N_{eu}})\to(\mathbb{Q},\sigma_{N_{eu}})$ be neutrosophic continuous and $g:(\mathbb{Q},\sigma_{N_{eu}})\to(\mathbb{R},\gamma_{N_{eu}})$ be perfectly $N_{eu}gs\alpha^*$ – continuous, then $gof:(\mathbb{P},\tau_{N_{eu}})\to(\mathbb{R},\gamma_{N_{eu}})$ is strongly $N_{eu}gs\alpha^*$ – continuous.

Proof:

Let A be any $N_{eu}gs\alpha^* - CS$ in $(\mathbb{R},\gamma_{N_{eu}})$. Since g is perfectly $N_{eu}gs\alpha^*$ – continuous, then $g^{-1}(A)$ is both $N_{eu} - OS$ and $N_{eu} - CS$ in $(\mathbb{Q},\sigma_{N_{eu}})$. Since f is neutrosophic continuous, then $f^{-1}(g^{-1}(A)) = (gof)^{-1}(A)$ is $N_{eu} - CS$ in $(\mathbb{P},\tau_{N_{eu}})$. Therefore, gof is strongly $N_{eu}gs\alpha^*$ – continuous.

Theorem 4.11: Let $f:(\mathbb{P},\tau_{N_{eu}})\to(\mathbb{Q},\sigma_{N_{eu}})$ be perfectly $N_{eu}gs\alpha^*$ – continuous and $g:(\mathbb{Q},\sigma_{N_{eu}})\to(\mathbb{R},\gamma_{N_{eu}})$ be strongly $N_{eu}gs\alpha^*$ – continuous, then $gof:(\mathbb{P},\tau_{N_{eu}})\to(\mathbb{R},\gamma_{N_{eu}})$ is perfectly $N_{eu}gs\alpha^*$ – continuous.

Proof:

Let A be any $N_{eu}gs\alpha^* - CS$ in $(\mathbb{R},\gamma_{N_{eu}})$. Since g is strongly $N_{eu}gs\alpha^*$ – continuous, then $g^{-1}(A)$ is $N_{eu} - CS$ in $(\mathbb{Q},\sigma_{N_{eu}}) \Rightarrow g^{-1}(A)$ is $N_{eu}gs\alpha^* - CS$ in $(\mathbb{Q},\sigma_{N_{eu}})$. Since f is perfectly $N_{eu}gs\alpha^*$ – continuous, then $f^{-1}(g^{-1}(A)) = (gof)^{-1}(A)$ is both $N_{eu} - OS$ and $N_{eu} - CS$ in $(\mathbb{P},\tau_{N_{eu}})$. Therefore, gof is perfectly $N_{eu}gs\alpha^*$ – continuous.

5. Totally neutrosophic $gs\alpha^*$ –continuous function

Definition 5.1: A neutrosophic function $f:(\mathbb{P},\tau_{N_{eu}})\to(\mathbb{Q},\sigma_{N_{eu}})$ is said to be totally $N_{eu}gs\alpha^*$ – continuous if the inverse image of every $N_{eu} - CS$ in $(\mathbb{Q},\sigma_{N_{eu}})$ is both $N_{eu}gs\alpha^* - OS$ and $N_{eu}gs\alpha^* - CS$ (ie, $N_{eu}gs\alpha^*$ – clopen set) in $(\mathbb{P},\tau_{N_{eu}})$.

Definition 5.2: A neutrosophic topological space $(\mathbb{P},\tau_{N_{eu}})$ is called a $N_{eu}gs\alpha^*$ –clopen set ($N_{eu}gs\alpha^*$ –clopen set) if it is both $N_{eu}gs\alpha^* - OS$ and $N_{eu}gs\alpha^* - CS$ in $(\mathbb{P},\tau_{N_{eu}})$.

Example 5.3: Let $\mathbb{P} = \{p\}$ and $\mathbb{Q} = \{q\}$. $\tau_{N_{eu}} = \{0_{N_{eu}}, 1_{N_{eu}}, A\}$ and $\sigma_{N_{eu}} = \{0_{N_{eu}}, 1_{N_{eu}}, B\}$ are N_{eu}TS on $(\mathbb{P},\tau_{N_{eu}})$ and $(\mathbb{Q},\sigma_{N_{eu}})$ respectively. Also $A = \{\langle p, (0.4, 0.5, 0.7)\rangle\}$ and $B = \{\langle q, (0.2, 0.7, 0.8)\rangle\}$ are $N_{eu}(\mathbb{P})$ and $N_{eu}(\mathbb{Q})$. Define a map $f:(\mathbb{P},\tau_{N_{eu}})\to(\mathbb{Q},\sigma_{N_{eu}})$ by $f(p) = q$. Let $B^c = \{\langle q, (0.8, 0.3, 0.2)\rangle\}$ be a $N_{eu} - CS$ in $(\mathbb{Q},\sigma_{N_{eu}})$. Then $f^{-1}(B^c) = \{\langle p, (0.8, 0.3, 0.2)\rangle\}$. $N_{eu}\alpha^* - OS = N_{eu}\alpha - OS = \{0_{N_{eu}}, 1_{N_{eu}}, A\}$ and $N_{eu}\alpha - CS = \{0_{N_{eu}}, 1_{N_{eu}}, A^c\}$.
$N_{eu}\alpha\text{-}cl\left(f^{-1}(B^c)\right) = 1_{N_{eu}}$. Now, $N_{eu}\alpha - int\left(N_{eu}\alpha - cl\left(f^{-1}(B^c)\right)\right) = 1_{N_{eu}} \subseteq N_{eu} - int(1_{N_{eu}}) = 1_{N_{eu}}$, whenever $f^{-1}(B^c) \subseteq 1_{N_{eu}} \Rightarrow f^{-1}(B^c)$ is $N_{eu}gs\alpha^* - CS$ in $(\mathbb{P},\tau_{N_{eu}})$.

Also, $N_{eu}\alpha\text{–}int\left(f^{-1}(\mathbb{B}^c)\right) = 0_{N_{eu}}$.Now, $N_{eu}\alpha - cl\left(N_{eu}\alpha\text{–}int\left(f^{-1}(\mathbb{B}^c)\right)\right) = 0_{N_{eu}} \supseteq N_{eu} - cl(0_{N_{eu}}) = 0_{N_{eu}}$, whenever $f^{-1}(\mathbb{B}^c) \supseteq 0_{N_{eu}} \Rightarrow f^{-1}(\mathbb{B}^c)$ is $N_{eu}gs\alpha^* - OS$ in $(\mathbb{P}, \tau_{N_{eu}})$. Therefore, f is totally $N_{eu}gs\alpha^*$ – continuous.

Theorem 5.4: Every perfectly $N_{eu}gs\alpha^*$ – continuous is totally $N_{eu}gs\alpha^*$ – continuous, but not conversely.

Proof:

Let $f : (\mathbb{P}, \tau_{N_{eu}}) \to (\mathbb{Q}, \sigma_{N_{eu}})$ be any neutrosophic function. Let $\mathbb{A}$ be any $N_{eu} - CS$ in $(\mathbb{Q}, \sigma_{N_{eu}})$. Then $\mathbb{A}$ is $N_{eu}gs\alpha^* - CS$ in $(\mathbb{Q}, \sigma_{N_{eu}})$. Since f is perfectly $N_{eu}gs\alpha^*$ – continuous, then $f^{-1}(\mathbb{A})$ is both $N_{eu} - OS$ and $N_{eu} - CS$ in $(\mathbb{P}, \tau_{N_{eu}}) \Rightarrow f^{-1}(\mathbb{A})$ is both $N_{eu}gs\alpha^* - OS$ and $N_{eu}gs\alpha^* - CS$ in $(\mathbb{P}, \tau_{N_{eu}})$. Therefore, f is totally $N_{eu}gs\alpha^*$ – continuous.

Example 5.5: Let $\mathbb{P} = \{p\}$ and $\mathbb{Q} = \{q\}$. $\tau_{N_{eu}} = \{0_{N_{eu}}, 1_{N_{eu}}, \mathbb{A}\}$ and $\sigma_{N_{eu}} = \{0_{N_{eu}}, 1_{N_{eu}}, \mathbb{B}\}$ are N_{eu}TS on $(\mathbb{P}, \tau_{N_{eu}})$ and $(\mathbb{Q}, \sigma_{N_{eu}})$ respectively. Also $\mathbb{A} = \{\langle p, (0.2, 0.4, 0.6)\rangle\}$ and $\mathbb{B} = \{\langle q, (0.6, 0.8, 0.4)\rangle\}$ are $N_{eu}(\mathbb{P})$ and $N_{eu}(\mathbb{Q})$. Define a map $f : (\mathbb{P}, \tau_{N_{eu}}) \to (\mathbb{Q}, \sigma_{N_{eu}})$ by $f(p) = q$. Let $\mathbb{B}^c = \{\langle q, (0.4, 0.2, 0.6)\rangle\}$ be a $N_{eu} - CS$ in $(\mathbb{Q}, \sigma_{N_{eu}})$. Then $f^{-1}(\mathbb{B}^c) = \{\langle p, (0.4, 0.2, 0.6)\rangle\}$. $N_{eu}\alpha^* - OS = N_{eu}\alpha - OS = \{0_{N_{eu}}, 1_{N_{eu}}, \mathbb{A}\}$ and $N_{eu}\alpha - CS = \{0_{N_{eu}}, 1_{N_{eu}}, \mathbb{A}^c\}$.
$N_{eu}\alpha\text{–}cl\left(f^{-1}(\mathbb{B}^c)\right) = \mathbb{A}^c \cap 1_{N_{eu}} = \mathbb{A}^c$.Now, $N_{eu}\alpha - int\left(N_{eu}\alpha - cl\left(f^{-1}(\mathbb{B}^c)\right)\right) = \mathbb{A} \cup 0_{N_{eu}} = \mathbb{A} \subseteq N_{eu} - int(1_{N_{eu}}) = 1_{N_{eu}}$, whenever $f^{-1}(\mathbb{B}^c) \subseteq 1_{N_{eu}} \Rightarrow f^{-1}(\mathbb{B}^c)$ is $N_{eu}gs\alpha^* - CS$ in $(\mathbb{P}, \tau_{N_{eu}})$. Also, $N_{eu}\alpha\text{–}int\left(f^{-1}(\mathbb{B}^c)\right) = 0_{N_{eu}}$. Now, $N_{eu}\alpha - cl\left(N_{eu}\alpha - int\left(f^{-1}(\mathbb{B}^c)\right)\right) = 0_{N_{eu}} \supseteq N_{eu} - cl(0_{N_{eu}}) = 0_{N_{eu}}$, whenever $f^{-1}(\mathbb{B}^c) \supseteq 0_{N_{eu}} \Rightarrow f^{-1}(\mathbb{B}^c)$ is $N_{eu}gs\alpha^* - OS$ in $(\mathbb{P}, \tau_{N_{eu}})$. Therefore, f is totally $N_{eu}gs\alpha^*$ – continuous. But f is not perfectly $N_{eu}gs\alpha^*$ – continuous. Let $\mathbb{T} = \{\langle q, (0.3, 0.1, 0.8)\rangle\}$ be $N_{eu}gs\alpha^* - CS$ in $(\mathbb{Q}, \sigma_{N_{eu}})$. Then $f^{-1}(\mathbb{T}) = \{\langle p, (0.3, 0.1, 0.8)\rangle\}$. Now, $N_{eu}\text{–}int\left(f^{-1}(\mathbb{T})\right) = 0_{N_{eu}} \neq f^{-1}(\mathbb{T}) \Rightarrow f^{-1}(\mathbb{T})$ is not $N_{eu} - OS$ in $(\mathbb{P}, \tau_{N_{eu}})$. Also, $N_{eu}\text{–}cl\left(f^{-1}(\mathbb{T})\right) = \mathbb{A}^c \neq f^{-1}(\mathbb{T}) \Rightarrow f^{-1}(\mathbb{T})$ is not $N_{eu} - CS$ in $(\mathbb{P}, \tau_{N_{eu}})$. Therefore, $f^{-1}(\mathbb{T})$ is not both $N_{eu} - OS$ and $N_{eu} - CS$ in $(\mathbb{P}, \tau_{N_{eu}})$.

Theorem 5.6: Every totally $N_{eu}gs\alpha^*$ – continuous is $N_{eu}gs\alpha^*$ – continuous.

Proof:

Let $f : (\mathbb{P}, \tau_{N_{eu}}) \to (\mathbb{Q}, \sigma_{N_{eu}})$ be any neutrosophic function. Let $\mathbb{A}$ be any $N_{eu} - CS$ in $(\mathbb{Q}, \sigma_{N_{eu}})$. Since f is totally $N_{eu}gs\alpha^*$ – continuous, then $f^{-1}(\mathbb{A})$ is both $N_{eu}gs\alpha^* - OS$ and $N_{eu}gs\alpha^* - CS$ in $(\mathbb{P}, \tau_{N_{eu}}) \Rightarrow f^{-1}(\mathbb{A})$ is $N_{eu}gs\alpha^* - CS$ in $(\mathbb{P}, \tau_{N_{eu}})$. Therefore, f is $N_{eu}gs\alpha^*$ – continuous.

Example 5.7: Let $\mathbb{P} = \{p\}$ and $\mathbb{Q} = \{q\}$. $\tau_{N_{eu}} = \{0_{N_{eu}}, 1_{N_{eu}}, \mathbb{A}\}$ and $\sigma_{N_{eu}} = \{0_{N_{eu}}, 1_{N_{eu}}, \mathbb{B}\}$ are N_{eu}TS on $(\mathbb{P}, \tau_{N_{eu}})$ and $(\mathbb{Q}, \sigma_{N_{eu}})$ respectively. Also $\mathbb{A} = \{\langle p, (0.7, 0.6, 0.5)\rangle\}$ and $\mathbb{B} = \{\langle q, (0.7, 0.8, 0.3)\rangle\}$ are $N_{eu}(\mathbb{P})$ and $N_{eu}(\mathbb{Q})$. Define a map $f : (\mathbb{P}, \tau_{N_{eu}}) \to (\mathbb{Q}, \sigma_{N_{eu}})$ by $f(p) = q$. Let $\mathbb{B}^c = \{\langle q, (0.3, 0.2, 0.7)\rangle\}$ be a $N_{eu} - CS$ in $(\mathbb{Q}, \sigma_{N_{eu}})$. Then $f^{-1}(\mathbb{B}^c) = \{\langle p, (0.3, 0.2, 0.7)\rangle\}$. $N_{eu}\alpha^* - OS = N_{eu}\alpha - OS = \{0_{N_{eu}}, 1_{N_{eu}}, \mathbb{A}, D\}$ and $N_{eu}\alpha - CS = \{0_{N_{eu}}, 1_{N_{eu}}, \mathbb{A}^c, E\}$, where $D = \{\langle p, ([0.7, 1], [0.6, 1], [0, 0.5])\rangle\}$, $E = \{\langle p, ([0, 0.5], [0, 0.4], [0.7, 1])\rangle\}$.
$N_{eu}\alpha\text{–}cl\left(f^{-1}(\mathbb{B}^c)\right) = \mathbb{A}^c \cap F \cap 1_{N_{eu}} = F$, where $= \{\langle p, ([0.3, 0.5], [0.2, 0.4], 0.7)\rangle\}$.
Now, $N_{eu}\alpha - int\left(N_{eu}\alpha - cl\left(f^{-1}(\mathbb{B}^c)\right)\right) = 0_{N_{eu}} \subseteq N_{eu} - int(\mathbb{A}), N_{eu} - int(D), N_{eu} - int(1_{N_{eu}}) = \mathbb{A}, 1_{N_{eu}}$, whenever $f^{-1}(\mathbb{B}^c) \subseteq \mathbb{A}, 1_{N_{eu}} \Rightarrow f^{-1}(\mathbb{B}^c)$ is $N_{eu}gs\alpha^* - CS$ in $(\mathbb{P}, \tau_{N_{eu}})$. Therefore, f is $N_{eu}gs\alpha^*$ – continuous. But f is not totally $N_{eu}gs\alpha^*$ – continuous, because $f^{-1}(\mathbb{B}^c)$ is not $N_{eu}gs\alpha^* - OS$ in $(\mathbb{P}, \tau_{N_{eu}})$.
Since $N_{eu}\alpha - cl\left(N_{eu}\alpha - int\left(f^{-1}(\mathbb{B}^c)\right)\right) = 0_{N_{eu}} \not\supseteq N_{eu} - cl(J) = \mathbb{A}^c$, whenever $f^{-1}(\mathbb{B}^c) \supseteq J$, where $J = \{\langle p, ([0, 0.3], [0, 0.2], [0.7, 1])\rangle\} \Rightarrow f^{-1}(\mathbb{B}^c)$ is not $N_{eu}gs\alpha^* - OS$ in $(\mathbb{P}, \tau_{N_{eu}})$.

Inter-relationship 5.8:

Perfectly $N_{eu}gs\alpha^*$ – continuous → Totally $N_{eu}gs\alpha^*$ – continuous

Totally $N_{eu}gs\alpha^*$ – continuous → $N_{eu}gs\alpha^*$ – continuous

Theorem 5.9: Let $f : (\mathbb{P}, \tau_{N_{eu}}) \to (\mathbb{Q}, \sigma_{N_{eu}})$ be totally $N_{eu}gs\alpha^*$ – continuous and $(\mathbb{Q}, \sigma_{N_{eu}})$ be $N_{eu}gs\alpha^* - T_{1/2}$ space, then f is $N_{eu}gs\alpha^*$ – irresolute.

Proof:

Let $\mathbb{A}$ be any $N_{eu}gs\alpha^* - CS$ in $(\mathbb{Q}, \sigma_{N_{eu}})$. Since $(\mathbb{Q}, \sigma_{N_{eu}})$ is $N_{eu}gs\alpha^* - T_{1/2}$ space, then $\mathbb{A}$ is $N_{eu} - CS$ in $(\mathbb{Q}, \sigma_{N_{eu}})$. Since f is totally $N_{eu}gs\alpha^*$ – continuous, then $f^{-1}(\mathbb{A})$ is both $N_{eu}gs\alpha^* - OS$ and $N_{eu}gs\alpha^* - CS$ in $(\mathbb{P}, \tau_{N_{eu}}) \Rightarrow f^{-1}(\mathbb{A})$ is $N_{eu}gs\alpha^* - CS$ in $(\mathbb{P}, \tau_{N_{eu}})$. Therefore, f is $N_{eu}gs\alpha^*$ – irresolute.

Theorem 5.10: Let $f : (\mathbb{P}, \tau_{N_{eu}}) \to (\mathbb{Q}, \sigma_{N_{eu}})$ and $g : (\mathbb{Q}, \sigma_{N_{eu}}) \to (\mathbb{R}, \gamma_{N_{eu}})$ be totally $N_{eu}gs\alpha^*$ – continuous and $(\mathbb{Q}, \sigma_{N_{eu}})$ be $N_{eu}gs\alpha^* - T_{1/2}$ space, then $gof : (\mathbb{P}, \tau_{N_{eu}}) \to (\mathbb{R}, \gamma_{N_{eu}})$ is totally $N_{eu}gs\alpha^*$ – continuous.

Proof:

Let $\mathbb{A}$ be any $N_{eu} - CS$ in $(\mathbb{R}, \gamma_{N_{eu}})$. Since g is totally $N_{eu}gs\alpha^*$ – continuous, then $g^{-1}(\mathbb{A})$ is both $N_{eu}gs\alpha^* - OS$ and $N_{eu}gs\alpha^* - CS$ in $(\mathbb{Q}, \sigma_{N_{eu}})$. Since $(\mathbb{Q}, \sigma_{N_{eu}})$ is $N_{eu}gs\alpha^* - T_{1/2}$ space, then $g^{-1}(\mathbb{A})$ is $N_{eu} - CS$ in $(\mathbb{Q}, \sigma_{N_{eu}})$. Since f is totally $N_{eu}gs\alpha^*$ – continuous, then $f^{-1}(g^{-1}(\mathbb{A})) = (gof)^{-1}(\mathbb{A})$ is both $N_{eu}gs\alpha^* - OS$ and $N_{eu}gs\alpha^* - CS$ in $(\mathbb{P}, \tau_{N_{eu}})$. Therefore, gof is totally $N_{eu}gs\alpha^*$ – continuous.

Author details

P. Anbarasi Rodrigo[1] and S. Maheswari[2*]

1 Assistant Professor, Department of Mathematics, St. Mary's College (Autonomous), Thoothukudi, Affiliated by Manonmaniam Sundaranar University, Tirunelveli, India

2 Research Scholar, Register number : 20212212092003, St. Mary's College (Autonomous), Thoothukudi, Affiliated by Manonmaniam Sundaranar University, Tirunelveli, India

*Address all correspondence to: mahma1295@gmail.com

References

[1] Floretin Smarandache, Neutrosophic Set: A Generalization of Intuitionistic Fuzzy Set, Jorunal of Defense Resources Management, 2010,107-116.

[2] Atanassov, K., Intuitionistic Fuzzy Sets, Fuzzy Sets And Systems, pp. 87-94, 1986.

[3] Zadeh, L.A., Fuzzy Sets, Inform And Control, Vol.8, pp.338-353, 1965.

[4] Dhavaseelan, R., & Jafari, S., Generalized Neutrosophic Closed Sets, New trends in Neutrosophic Theory And Applications, Vol.2, pp.261-273, Jan 2017.

[5] Sreeja, D., & Sarankumar, T., Generalized Alpha Closed Sets in Neutrosophic Topological Spaces, JASC, Journal of Applied Science & Computations, ISSN:1076-5131, Vol.5, Issue 11, pp.1816-1823, Nov-2018.

[6] Anbarasi Rodrigo, P., Maheswari, S., Neutrosophic Generalized Semi Alpha Star Closed Sets in Neutrosophic Topological Spaces, Paper presented in International Conference on Mathematics, Statistics, Computers And Information Sciences, 2021.

[7] Anbarasi Rodrigo, P., Maheswari, S., Functions Related to Neutrosophic $gs\alpha^{*}-$ Closed Sets in Neutrosophic Topological Spaces, Paper presented in 24th FAI International Conference on Global Trends of Data Analytics in Business Management, Social Sciences, Medical Sciences and Decision Making 24th FAI-ICDBSMD 2021.

[8] Blessie Rebecca, S., Francina Shalini, A., Neutrosophic Generalized Regular Contra Continuity in Neutrosophic Topological Spaces, Research in Advent Technology, Vol.7, No.2, E-ISSN: 2321-9637, pp.761-765, Feb 2019.

[9] Serkan Karatas, Cemil Kuru, Neutrosophic Topology, Neutrosophic Sets & Systems, University of New Mexico, Vol.13, pp.90-95, 2016.

Chapter 3

4-Dimensional Canards with Brownian Motion

Shuya Kanagawa and Kiyoyuki Tchizawa

Abstract

Generally speaking, it is impossible to analyze slow-fast system with Brownian motion. If it becomes possible to do using a new approach, we can evaluate the rigidity of the original system. What kind of the behavior of such a system we have? Using non-standard analysis, on a“hyper finite time line” by Anderson, the Brownian motions are described by step functions. Then, the original differential equations are described by the difference equations due to using non-standard analysis. When constructing the difference equations, the corresponding measure is extended topologically. Because the interval of the difference is according to the hyper finite time line, the topological space is well defined. In this paper, we propose a two-region economic model with Brownian motions. This concrete example gives us new results.

Keywords: canard solution, slow-fast system, nonstandard analysis, Brownian motion, stochastic differential equation

1. Introduction

Consider a slow-fast system in $\mathbf{R}^4$ with a 2-dimensional slow manifold:

$$\begin{cases} \varepsilon \dfrac{dx}{dt} = h(x, y, \varepsilon) \\ \dfrac{dy}{dt} = g(x, y, \varepsilon) \end{cases}, \qquad (1)$$

where ε is infinitesimal and

$$g = \begin{pmatrix} g_1 \\ g_2 \end{pmatrix} : \mathbf{R}^4 \to \mathbf{R}^2, \ \ h = \begin{pmatrix} h_1 \\ h_2 \end{pmatrix} : \mathbf{R}^4 \to \mathbf{R}^2,$$
$$x = x(t) = \begin{pmatrix} x_1 \\ x_2 \end{pmatrix} \in \mathbf{R}^2, \ \ y = y(t) = \begin{pmatrix} y_1 \\ y_2 \end{pmatrix} \in \mathbf{R}^2.$$

The above expression form is based on Nelson’s [1]. The slow-fast system (1) is applied to many fields, e.g. electronic circuits, neuron systems, etc. In these applications, effectiveness of random noises always exists. We take up this point of view as being one of the main problems.

Now, let us consider a stochastic differential equation for a slow-fast system with a Brownian motion $B(t)$ as the random noises modifying the slow fast system (1): For $t \in [0, T], \quad T > 0$

$$\begin{cases} \varepsilon dx = h(x, y, \varepsilon)dt \\ dy = g(x, y, \varepsilon)dt + \sigma dB, \end{cases} \tag{2}$$

where $B = \begin{pmatrix} B_1 \\ B_2 \end{pmatrix} \in R^2$ is a 2-dimensional standard Brownian motion and $\sigma > 0$ is a positive constant which gives a standard deviation for the Brownian motion $B(t)$.

Since the Brownian motion $B(t)$ is almost surely non-differentiable everywhere, it is difficult to analyze slow-fast system (2).

On the other hand, Anderson [2] showed that the Brownian motion is described by step functions using non-standard analysis on a hyper finite time line by the following definition. (See also [3, 4]).

Definition 1. *Let* $N_t = \frac{t}{\Delta t}, \quad 0 \le t \le T$ *and* $N = N_T$. *Assume that a sequence of i.i.d. random variables* $\{\Delta B_k, \quad k = 1, \cdots, N\}$ *has the distribution*

$$P\left\{\Delta B_k = \sqrt{\Delta t}\right\} = P\left\{\Delta B_k = -\sqrt{\Delta t}\right\} = \frac{1}{2} \tag{3}$$

for each $k = 1, \cdots, N$. An extended Wiener process $\{B(t), \quad t \ge 0\}$ is defined by

$$B(t) = \sum_{k=1}^{Nt} \Delta B_k, \quad 0 \le t \le T. \tag{4}$$

Rewriting the system (2) via step functions on the hyper finite time line, the following system (5) is obtained.

$$\begin{cases} \varepsilon(x_n - x_{n-1}) = h(x_{n-1}, y_{n-1}, \varepsilon)\Delta t \\ y_n - y_{n-1} = g(x_{n-1}, y_{n-1}, \varepsilon)\Delta t + \sigma \Delta B_n, \end{cases} \tag{5}$$

for $n = 1, 2, \quad \cdots \quad, N$, where $\Delta B_n = B(n\Delta t) - B((n-1)\Delta t)$, $\Delta t = \frac{T}{N}$ and N is a hyper finite natural number.

Since the system (5) is equivalent to the system (2), taking B(t) in Definition 1, we prove the existence of the solution for the system (2) in Section 3.

2. Slow-fast system in $\mathbf{R}^4$ with co-dimension 2

We assume that the system (1) sastisfies the following conditions (A1) $\sim$ (A5):

(A1) h is of class $\mathbf{C}^1$ and g is of class $\mathbf{C}^2$.

(A2) The slow manifold $S = \{(x, y) \in \mathbf{R}^4 | h(x, y, 0) = 0\}$ is a two-dimensional differential manifold and intersects the set $V = \{(x, y) \in \mathbf{R}^4 | \det(\frac{\partial h}{\partial x}(x, y, 0)) = 0\}$ transversely.

Then, the pli set $PL = \{(x, y) \in S \cap V\}$ is a one-dimensional differentiable manifold.

(A3) Either the value of g_1 or that of g_2 is nonzero at any point of PL.

Note that the pli set PL devides the slow manifolds $S \backslash \mathrm{PL}$ into three parts depending on the signs of the two eigenvalues of $\frac{\partial h}{\partial x}(x, y, 0)$.

First consider the following reduced system which is obtained from (1) with $\varepsilon = 0$:

$$\begin{cases} 0 = h(x, y, 0) \\ \dfrac{dy}{dt} = g(x, y, 0). \end{cases} \tag{6}$$

By differentiating $h(x, y, 0)$ with respect to t, we have

$$\frac{\partial h}{\partial x}(x, y, 0)\frac{dx}{dt} + \frac{\partial h}{\partial y}(x, y, 0)g(x, y, 0) = 0. \tag{7}$$

Then (6) becomes the following:

$$\begin{cases} \dfrac{dx}{dt} = -\left[\dfrac{\partial h}{\partial x}(x, y, 0)\right]^{-1}\dfrac{\partial h}{\partial y}(x, y, 0)g(x, y, 0) \\ \dfrac{dy}{dt} = g(x, y, 0) \end{cases}, \tag{8}$$

where $(x, y) \in S \backslash PL$. To avoid degeneracy in (8), we consider the following system:

$$\begin{cases} \dfrac{dx}{dt} = \left\{-\det\left[\dfrac{\partial h}{\partial x}(x, y, 0)\right]^{-1}\right\}\left[\dfrac{\partial h}{\partial x}(x, y, 0)\right]^{-1}\dfrac{\partial h}{\partial y}(x, y, 0)g(x, y, 0) \\ \dfrac{dy}{dt} = \left\{\det\left[\dfrac{\partial h}{\partial x}(x, y, 0)\right]^{-1}\right\}g(x, y, 0) \end{cases}. \tag{9}$$

The phase portrait of the system (9) is the same as that of (8) except the region where $\det\left[\frac{\partial h}{\partial x}(x, y, 0)\right] = 0$, but only the orientation of the orbit is different.

Definition 2. A singular point of (9), which is on PL, is called a pseudo singular point of (1).

(A4) $rank\left[\frac{\partial h}{\partial x}(x, y, 0)\right] = 2$ for any $(x, y) \in S \backslash \mathrm{PL}$.

From (A4), the implicit function theorem guarantees the existence of a unique function $y = \xi(x)$ such that $h(x, \xi(x), 0) = 0$. By using $y = \xi(x)$, we obtain the following system:

$$\frac{dx}{dt} = \left\{-\det\left[\frac{\partial h}{\partial x}(x, \xi(x), 0)\right]^{-1}\right\}\left[\frac{\partial h}{\partial x}(x, \xi(x), 0)\right]^{-1}\frac{\partial h}{\partial y}(x, \xi(x), 0)g(x, \xi(x), 0). \tag{10}$$

(A5) All singular points of (10) are non-degenerate, that is, the linearization of (10) at a singular point has two nonzero eigenvalues.

Definition 3. Let λ_1, λ_2 be two eigenvalues of the linearization of (10) at a pseudo singular point. The pseudo singular point with real eigenvalues is called a pseudo singular saddle point if $\lambda_1 < 0 < \lambda_2$ and a pseudo singular node point if $\lambda_1 < \lambda_2 < 0$ or $\lambda_1 > \lambda_2 > 0$.

The following theorem is established (see, e.g. [5]).

Theorem 1. Let (x_0, y_0) be a pseudo singular point. If $\mathrm{trace}\left[\frac{\partial h}{\partial x}(x_0, y_0, 0)\right] < 0$, then there exists a solution which first follows the attractive part and the repulsive part after crossing PL near the pseudo singular point.

Remark 1. The condition $\mathrm{trace}\left[\frac{\partial h}{\partial x}(x_0, y_0, 0)\right] < 0$ implies that one of eigenvalues of $\left[\frac{\partial h}{\partial x}(x_0, y_0, 0)\right]$ is equal to zero and the other one is negative. Notice that the system has two kinds of vector fields: one is 2-dimensional slow and the other is 2-dimensional fast one. The condition provides the state of the fast vector field.

Remark 2. The singular solution in Theorem 1 is called a canard in $\mathbf{R}^4$ with 2-dimensional slow manifold. As a result, it causes a delayed jumping. The study of canards requires still more precise topological analysis on the slow vector field.

In the next section, we show that a canard exists for the system (2) in which the orbit of the canard of the system (1) is moved to another one by a Brownian motion $B(t)$.

3. Canards with Brownian motion

Let us prove the following theorem.

Theorem 2. *In the system (3), if there exists* $\{k_n\}$ *such that*

$$|x_n - x_{n-1}| \leq \varepsilon k_n, \quad n = 1, 2, \cdots, N \tag{11}$$

and

$$\sup_{1 \leq n \leq N} k_n \leq K \tag{12}$$

for some K hyper finite, then there exists a solution of (5) which is called canard in the sense of Remark 2.

Proof. From the condition (11), we have

$$|x_n - x_{n-1}| = \left| \frac{1}{\varepsilon} h\left(x_{n-1}, y_{n-1}, \varepsilon\right) \Delta t \right| \leq \varepsilon k_n. \tag{13}$$

ε is an arbitrary constant, therefore putting $\varepsilon = \frac{1}{N}$ we have from (13)

$$\left| h\left(x_{n-1}, y_{n-1}, \varepsilon\right) \right| \leq \frac{\varepsilon^2 k_n}{\Delta t} \leq \frac{k_n}{N}, \tag{14}$$

for each $n \geq 1$.

From Definition 2, the following is satisfied for the pseudo-singular point (x_0, y_0) of (1);

$$\begin{cases} \left\{ -\det\left[\frac{\partial h}{\partial x}\left(x_0, y_0, 0\right)\right]^{-1} \right\} \left[\frac{\partial h}{\partial x}\left(x_0, y_0, 0\right)\right]^{-1} \dfrac{\partial h}{\partial y}\left(x_0, y_0, 0\right) g\left(x_0, y_0, 0\right) = 0 \\ h\left(x_0, y_0, 0\right) = 0 \end{cases} . \tag{15}$$

Assume that σ^2 in the Brownian motion $B(t)$ is sufficiently small. Let $\left(x_\xi, y_\xi\right)$ be a pseudo-singular point of (2) or (5). Note that (2) is equivalent to (5) in the sense of nonstandard analysis. Then there exists a positive number ξ such that

$$\left\{ -\det\left[\frac{\partial h}{\partial x}\left(x_\xi, y_\xi, 0\right)\right]^{-1} \right\} \left[\frac{\partial h}{\partial x}\left(x_\xi, y_\xi, 0\right)\right]^{-1} \frac{\partial h}{\partial y}\left(x_\xi, y_\xi, 0\right) \left(g\left(x_\xi, y_\xi, 0\right) \Delta t + \sigma \Delta B_\xi \right) \approx 0, \tag{16}$$

where $\Delta t = \frac{T}{N}$.

In this situation, as $\sigma \approx 0$

$$\left(x_\xi, y_\xi\right) \approx \left(x_0, y_0\right). \tag{17}$$

Therefore, the eigenvalues of the linearized system (2) at the point $\left(x_\xi, y_\xi\right)$ keeps the almost same as the eigenvalues of the system (1) at the point (x_0, y_0).

On the other hand, there exists a canard of (1) from Theorem 1. Since $\frac{k_n}{N}$ is small enough, the solution of (5) also first follows the attractive part and the repulsive part follows after crossing PL near the pseudo singular point like as the canard of (1).

4. Concrete models

4.1 Two-region business cycle model

As a concrete model, we consider a two-region business cycle model between two nations A and B including a Brownian motion $B(t) = (B_1(t), B_2(t))$ as followings. See [6] for more details of the two-region business cycle model.

$$\begin{cases} \varepsilon \dfrac{dx_1}{dt} = -\dfrac{1-\alpha+m_1}{\theta} y_1 + \dfrac{m_2}{\theta} y_2 - \left(\dfrac{\varepsilon}{\theta} + 1 - \alpha\right)(x_1 - a) + \dfrac{1-n_1}{\theta}\varphi_1(x_1 - a) + \dfrac{n_2}{\theta}\varphi_2(x_2 - a) \\ \varepsilon \dfrac{dx_2}{dt} = \dfrac{m_1}{\theta} y_1 - \dfrac{1-\alpha+m_2}{\theta} y_2 + \dfrac{1-n_1}{\theta}\varphi_1(x_1 - a) - \left(\dfrac{\varepsilon}{\theta} + 1 - \alpha\right)(x_2 - a) + \dfrac{1-n_2}{\theta}\varphi_2(x_2 - a) \\ dy_1 = (x_1 - a)dt + \sigma dB_1(t) \\ dy_2 = (x_2 - a)dt + \sigma dB_2(t) \end{cases} \tag{18}$$

for $0 \le t \le T$, where $x_1(t)$ and $x_2(t)$ are exports of A and B, $m_1(t)$ and $m_2(t)$ are imports of A and B, $y_1(t) - \frac{q}{1-\alpha}$ and $y_2(t) - \frac{q}{1-\alpha}$ are national income identities of A and B for some constants q and α, respectively. (See [6] for more details.)

Now, let us introduce a difference equations for the system (18). Then, the relations $\Delta t = \frac{T}{N}$ and $t_k = k\frac{T}{N}$, $k = 0, 1, \cdots, N$ are satisfied, where N is a hyper finite. Put

$$\begin{cases} \dfrac{\varepsilon}{\Delta t}\{x_1(t_k) - x_1(t_{k-1})\} = -\dfrac{1-\alpha+m_1}{\theta} y_1(t_{k-1}) + \dfrac{m_2}{\theta} y_2(t_{k-1}) \\ \qquad -\left(\dfrac{\varepsilon}{\theta} + 1 - \alpha\right)(x_1(t_{k-1}) - a) + \dfrac{1-n_1}{\theta}\varphi_1(x_1(t_{k-1}) - a) + \dfrac{n_2}{\theta}\varphi_2(x_2(t_{k-1}) - a) \\ \dfrac{\varepsilon}{\Delta t}\{x_2(t_k) - x_2(t_{k-1})\} = \dfrac{m_1}{\theta} y_1(t_{k-1}) - \dfrac{1-\alpha+m_2}{\theta} y_2(t_{k-1}) \\ \qquad + \dfrac{1-n_1}{\theta}\varphi_1(x_1(t_{k-1}) - a) - \left(\dfrac{\varepsilon}{\theta} + 1 - \alpha\right)(x_2(t_{k-1}) - a) + \dfrac{1-n_2}{\theta}\varphi_2(x_2(t_{k-1}) - a) \\ y_1(t_k) - y_1(t_{k-1}) = (x_1(t_{k-1}) - a)\Delta t + \sigma\{B_1(t_k) - B_1(t_{k-1})\} \\ y_2(t_k) - y_2(t_{k-1}) = (x_2(t_{k-1}) - a)\Delta t + \sigma\{B_2(t_k) - B_2(t_{k-1})\} \end{cases} \tag{19}$$

Furthermore put

$$\varphi_1(x - \alpha) = \varphi_2(x - \alpha) = (1 - \alpha)\left(\theta x + x^2 - \frac{x^3}{3}\right). \tag{20}$$

4.2 Simulation results

In this section, let us provide computer simulations for the two-region business cycle model using the above Eqs. (19) and (20. In (19), we assume that two Brownian motions $B_1(t)$ and $B_2(t)$ are mutually independent and note that

$$B_1(t_k) - B_1(t_{k-1}) \sim N\left(0, \Delta t\sigma_1^2\right), \quad B_2(t_k) - B_2(t_{k-1}) \sim N\left(0, \Delta t\sigma_2^2\right), \tag{21}$$

for each $1 \le k \le \frac{T}{\Delta t}$.

Figure 1, except for the axes, shows the pli set $PL = \{(x, y) \in S \cap V\}$ with the pseudo singular point $(1, -1)$ of (1) defined by (9).

Putting some parameters in (19), we have the following results for some orbits of $\{(x_1(t), x_2(t)), \ \ 0 \le t \le T\}$ satisfying the Eq. (1) or (2).

Figure 2 shows an orbit of $\{(x_1(t), x_2(t)), \ \ 0 \le t \le T = 0.8\}$ satisfying the Eq. (5) with $\sigma_1 = \sigma_2 = 0$ and starting from $(0.8, -0.8)$ near the pseudo singular point $(1, -1)$. In **Figures 2–4**, $\varepsilon = 0.01$. From **Figure 2** the speed of the orbit $(x_1(t), x_2(t))$ for $0 \le t \le 0.2$ is not only very fast, but also the orbit jumps near the pseudo singular point $(1, -1)$. The orbit turns at the point $(-2, 2)$ and returns on the line. $\Delta t = 0.001$ in **Figures 2–6**.

Figure 3 shows an orbit of $\{(x_1(t), x_2(t)), \ \ 0 \le t \le T = 0.8\}$ satisfying the Eq. (5) with $\sigma_1 = \sigma_2 = 0.4$ and starting from $(0.8, -0.8)$ near the the pseudo singular point $(1, -1)$. From **Figure 3** we observe that the orbit moves on the line from $(0.8, -0.8)$ and separates from the line at $t = 0.2$ by the Brownian motion $B(t)$.

Figure 4 shows an orbit of $\{(x_1(t), x_2(t)), \ \ 0 \le t \le T = 4.64\}$ satisfying the Eq. (5) with $\sigma_1 = \sigma_2 = 0$ and starting from $(0.8, -0.8)$ near the pseudo singular point $(1, -1)$. The orbit separates from the line at $t = 4.61$.

Figure 5 shows an orbit of $\{(x_1(t), x_2(t)), \ \ 0 \le t \le T = 2.75\}$ satisfying the Eq. (5) with $\sigma_1 = \sigma_2 = 0$ and starting from $(0.8, -0.8)$ near the pseudo singular point $(1, -1)$, where $\varepsilon = 0.004$. The orbit with $\varepsilon = 0.004$ separates from the line at

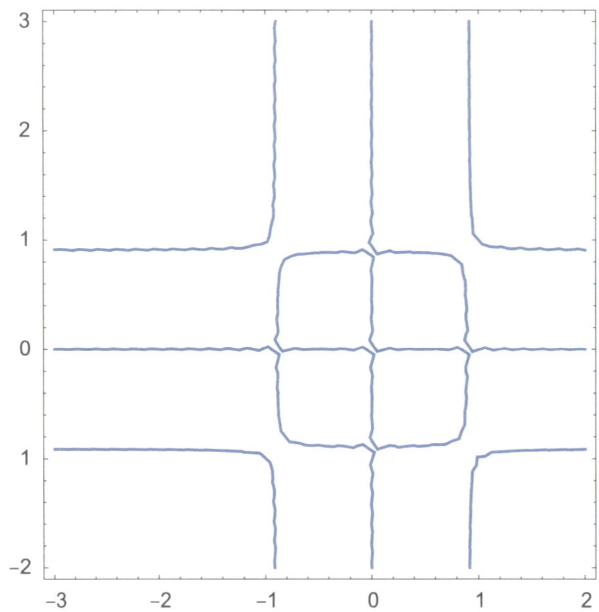

Figure 1.
Pli set PL.

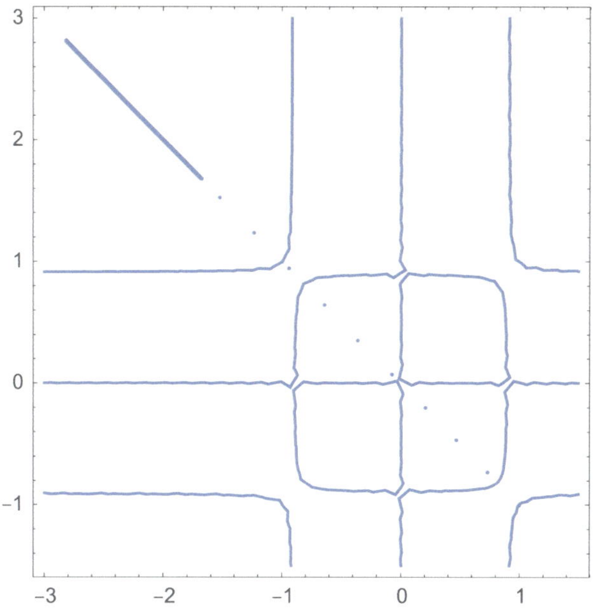

Figure 2.
$\Delta t = 0.001,\ 0 \le t \le 0.8,\ a = 0.6,\ \varepsilon = 0.01,\ \theta = 0.5, m_1 m_1 = 0.1,\ m_2 = 0.2,\ n_1 = 0.2,\ n_2 = 0.2,$
$\sigma_1 = \sigma_2 = 0.$

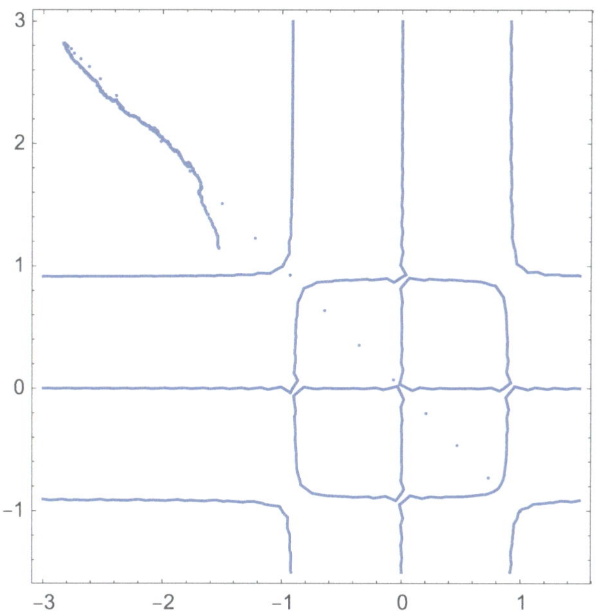

Figure 3.
$\Delta t = 0.001,\ 0 \le t \le 0.8,\ a = 0.6,\ \varepsilon = 0.01,\ \theta = 0.5, m_1 = 0.1,\ m_2 = 0.2,\ n_1 = 0.4,\ n_2 = 0.4,$
$\sigma_1 = \sigma_2 = 0.4.$

$t = 2.61$. On the other hand the orbit with $\varepsilon = 0.01$ separates from the line at $t = 4.61$ in **Figure 4**. Therefore we see that the orbit changes according to ε.

Figure 6 shows an orbit of $\{(x_1(t), x_2(t)),\ \ 0 \le t \le T = 0.8\}$ satisfying the Eq. (5) with $\sigma_1 = \sigma_2 = 0.4$ and starting from $(0.8, -0.8)$ near the pseudo singular point $(1, -1)$, where $\varepsilon = 0.004$. The orbit with $\varepsilon = 0.004$ separates from

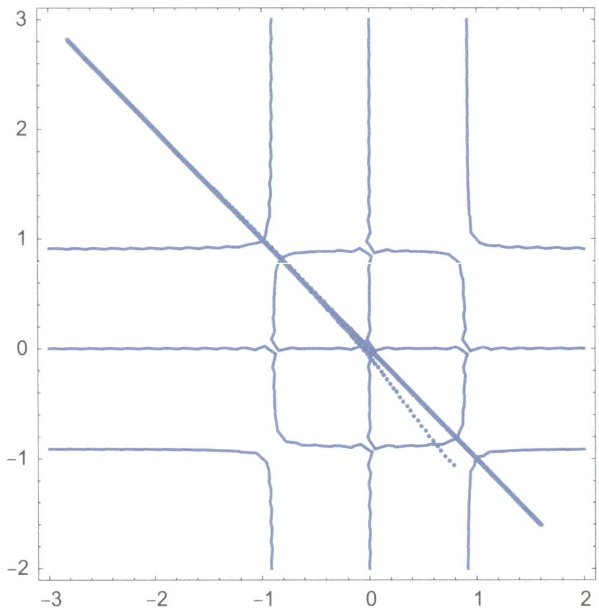

Figure 4.
$\Delta t = 0.001$, $0 \leq t \leq 4.64$, $a = 0.6$, $\varepsilon = 0.01$, $\theta = 0.5$, $m_1 = 0.1$, $m_2 = 0.2$, $n_1 = 0.2$, $n_2 = 0.2$, $\sigma_1 = \sigma_2 = 0$.

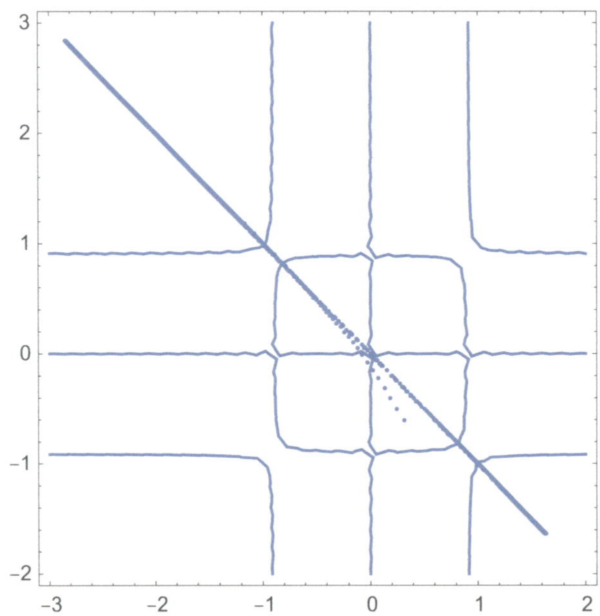

Figure 5.
$\Delta t = 0.001$, $0 \leq t \leq 2.75$, $a = 0.6$, $\varepsilon = 0.004$, $\theta = 0.5$, $m_1 = 0.1$, $m_2 = 0.2$, $n_1 = 0.2$, $n_2 = 0.2$, $\sigma_1 = \sigma_2 = 0$.

the line at $t = 0.1$. On the other hand, in **Figure 3**, the orbit with $\varepsilon = 0.01$ separates from the line at $t = 0.2$. Then, the orbit changes according to ε also in the non-random case.

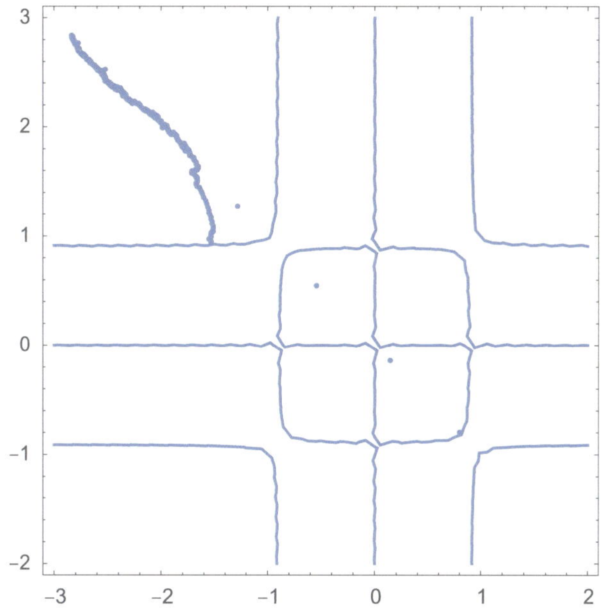

Figure 6.
$\Delta t = 0.001$, $0 \leq t \leq 0.8$, $a = 0.6$, $\varepsilon = 0.004$, $\theta = 0.5, m_1 = 0.1$, $m_2 = 0.2$, $n_1 = 0.2$, $n_2 = 0.2$, $\sigma_1 = \sigma_2 = 0.4$.

5. Conclusion

Brownian motions are described by non-differentiable functions almost surely. In order to overcome the difficulty in the system (2) we consider the system (5) using nonstandard analysis. The system (5) makes us possible to analyze the canard with Brownian motions. As the difference equations is determined by according to the hyper finite time line, the measure is extended effectively to do this analysis. In **Figures 1–6** obtained by the simulations, we observe the effects of Brownian motions which change the orbit of $(x_1(t), x_2(t))$.

Acknowledgements

The authors would like to express the reviewer's comments which are useful to explain the structure of canards. The first author is supported in part by Grant-in-Aid Scientific Research (C), No.18 K03431, Ministry of Education, Science and Culture, Japan.

The 2020 AMS classification

ordinary differential Eqs., dynamical systems and ergodic theory, difference and functional equations.

Author details

Shuya Kanagawa[1*] and Kiyoyuki Tchizawa[2]

1 Deparatment of Mathematics, Tokyo Gakugei University, Tokyo, Japan

2 Institute of Administration Engineering Ltd., Tokyo, Japan

*Address all correspondence to: kanagawa1954@icloud.com

References

[1] Nelson E. Internal set theory: A new approach to nonstandard analysis. Bulletin of the American Mathematical Society. 1977;**83**:1165-1198

[2] Anderson RM. A non-standard representation for brownian motion and ito integration. Israel Journal of Mathematics. 1976;**25**(2):1546

[3] Kanagawa S, Tchizawa K. Proof of Ito's Formula for Ito's Process in Nonstandard Analysis. Applied Mathematics. 2019;**10**(2):561-567

[4] Kanagawa S, Tchizawa K. Extended wiener process in nonstandard analysis. Applied Mathematics. 2020;**11**(2): 247254

[5] Tchizawa K. Four-dimensional canards and their center manifold. Extended Abstracts Spring 2018. Trends in Mathematics. 2019;**11**:193-199. Springer Nature Switzerland AG

[6] Miki H, Nishino H, Tchizawa K. On the possible occurrence of duck solutions in domestic and two-region business cycle models. Nonlinear Studies. 2012;**18**:39-55

Section 3

Cobordisms, Coverings and Topological Sheepers

Chapter 4

The Topology of the Configuration Space of a Mathematical Model for Cycloalkenes

Yasuhiko Kamiyama

Abstract

As a mathematical model for cycloalkenes, we consider equilateral polygons whose interior angles are the same except for those of the both ends of the specified edge. We study the configuration space of such polygons. It is known that for some case, the space is homeomorphic to a sphere. The purpose of this chapter is three-fold: First, using the h-cobordism theorem, we prove that the above homeomorphism is in fact a diffeomorphism. Second, we study the best possible condition for the space to be a sphere. At present, only a sphere appears as a topological type of the space. Then our third purpose is to show the case when a closed surface of positive genus appears as a topological type.

Keywords: cycloalkene, polygon, configuration space, h-cobordism theorem, closed surface

1. Introduction

The configuration space of mechanical linkages in the Euclidean space of dimension three, also known as polygon space, is the central objective in topological robotics. The linkage consists of n bars of length $l_1, \cdots, l_n$ connected by revolving joints forming a closed spatial polygonal chain.

The polygon space is quite important in various engineering applications: In molecular biology they describe varieties of molecular shapes, in robotics they appear as spaces of all possible configurations of some mechanisms, and they play a central role in statistical shape theory.

Mathematically, these spaces are also very interesting: The symplectic structure on the polygon space was studied in the seminal paper [1]. The integral cohomology ring was determined in [2] applying methods of toric topology. We refer to [3] for an excellent exposition with emphasis on Morse theory.

Recently, mathematicians are interested in a mathematical model for monocyclic hydrocarbons. The model is defined by imposing conditions on the interior angles of a polygon. The configuration space of such polygons corresponds in chemistry to the conformations of all possible shapes of a monocyclic hydrocarbon. Hence the configuration space is interesting both in mathematics and chemistry.

In order to give more detailed account, recall that monocyclic hydrocarbons are classified into two types: One is saturated type, and the other is unsaturated type. Mathematicians constructed a mathematical model for each type. We summarize the correspondence between chemical and mathematical terminologies in the following **Table 1**.

Below we explain **Table 1**.

i. Monocyclic saturated hydrocarbons are monocyclic hydrocarbons that contain only single bonds between carbon atoms. Monocyclic saturated hydrocarbons are called *cycloalkanes*. (See **Figure 1** for the 6-membered cycloalkane.)

- The mathematical model for cycloalkanes is the equilateral and equiangular polygons. Let $\mathcal{M}_n(\theta)$ be the configurations of such n-gons with interior angle θ. The study of the topological type of $\mathcal{M}_n(\theta)$ originated in [4]. See the next item for more details.

- The topological type of $\mathcal{M}_4(\theta)$ and $\mathcal{M}_5(\theta)$ was determined in [4] for arbitrary θ, and that of $\mathcal{M}_6(\theta)$ was determined in [5] for arbitrary θ. The paper [4] also determined the topological type of $\mathcal{M}_7(\theta)$ for the case that θ is the ideal tetrahedral bond angle, i.e. $\theta = \arccos\left(-\frac{1}{3}\right) \approx 109.47°$. The result was generalized in [6] for generic θ.

ii. Monocyclic unsaturated hydrocarbons are monocyclic hydrocarbons with at least one double or triple bond between carbon atoms.

- Hereafter, for simplicity, we consider only the monocyclic unsaturated hydrocarbons that contain exactly one multiple bond.

- It is not mathematically important whether the multiple bond is a double or triple bond. Hence we assume that the multiple bond is a double bond.

Chemistry	Mathematics
Monocyclic hydrocarbon	Equilateral polygon in $\mathbb{R}^3$
Bond of a monocyclic hydrocarbon	Edge of a polygon
Bond angle of a monocyclic hydrocarbon	Interior angle of an equilateral polygon
Conformations	Configuration space
Cycloalkanes	Equilateral and equiangular polygons. Their configuration space is denoted by $\mathcal{M}_n(\theta)$.
Cycloalkenes	Equilateral polygons whose interior angles are the same except for those of the both ends of the specified edge. Their configuration space is denoted by $C_n(\theta)$.

Table 1.
The correspondence between chemical and mathematical terminologies.

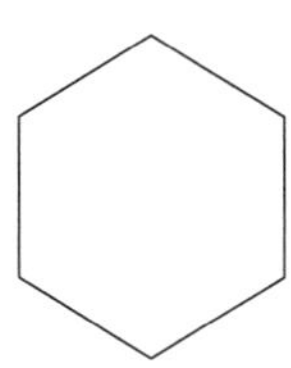

Figure 1.
Cyclohexane (6-membered cycloalkane).

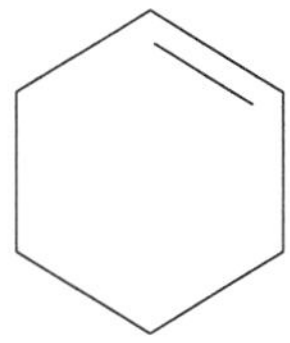

Figure 2.
Cyclohexene (6-membered cycloalkene).

- Such monocyclic unsaturated hydrocarbons are called *cycloalkenes*. (See **Figure 2** for the 6-membered cycloalkene.)

- The mathematical model for cycloalkenes is the equilateral polygons whose interior angles are the same except for those of the both sides of the specified edge. Here the specified edge corresponds to the double bond. Let $C_n(\theta)$ be the configurations of such n-gons with interior angle θ. (See (1) for more precise definition of $C_n(\theta)$.) The study of the topological type of $C_n(\theta)$ originated in [7] and the result was generalized in [8]. See the next item for more details.

- The following result was proved in [8] (see Theorem 6): There exists θ_0 such that for all $\theta \in \left(\theta_0, \frac{n-2}{n}\pi\right)$, $C_n(\theta)$ is *homeomorphic* to S^{n-4}.

- Except for the above result in [8], we do not know strong results about the topology of $C_n(\theta)$.

On the other hand, as a combinatorial result, the necessary and sufficient condition for $\mathcal{M}_n(\theta)$ and $C_n(\theta)$ to be non-empty was proved in [9]. (See Theorem 3 about the result for $C_n(\theta)$.)

As stated in the last item of the above ii, we do not have enough information about the topology of $C_n(\theta)$. The purpose of this chapter is to obtain systematic information about $C_n(\theta)$. More precisely, we study the following:

Problem 1. (i) We prove that the above homeomorphism in [8] is in fact a diffeomorphism.

(ii) We study the best possible value about the above θ_0 in [8].

(iii) At present, only a sphere appears as a topological type of $C_n(\theta)$. We determine the topological type of $C_6(\theta)$ for all θ. The result shows that for some θ, $C_6(\theta)$ is a closed surface of positive genus.

This chapter is organized as follows. In §2, we state our main results. In §3-§5, we prove them. In §6, we state the conclusions.

2. Main results

We give the definition of the configuration space. Let θ be a real number satisfying $0 \leq \theta \leq \pi$. We set

$$C_n(\theta) := \left\{P = (u_1, \cdots, u_n) \in \left(S^2\right)^n \mid \text{the following i, ii and iii hold}\right\}. \qquad (1)$$

i. $u_1 = (1, 0, 0)$ and $u_n = (-\cos\theta, -\sin\theta, 0)$.

ii. $\sum_{i=1}^{n} u_i = 0$.

iii. $\langle u_i, u_{i+1}\rangle = -\cos\theta$ for $1 \le i \le n-3$, where $\langle\ ,\rangle$ denotes the standard inner product on $\mathbb{R}^3$.

About the conditions in (1), the following explanations are in order. (See **Table 1** for chemical terminologies.)

- The element u_i denotes the unit vector in the direction of the edge of a polygon. Then the condition ii requires the fact that $(u_1, \cdots, u_n)$ is in fact a polygon.

- We specify u_{n-1} to be the special edge, which corresponds to the double bond of a cycloalkene. Then the condition iii requires the fact that the interior angles of an n-gon are θ except for those of the both ends of u_{n-1}.

Remark 2. In some papers, $C_n(\theta)$ is defined as

$$A_n(\theta)/SO(3), \tag{2}$$

where we set

$$A_n(\theta) := \left\{(u_1, \cdots, u_n) \in \left(S^2\right)^n \mid (1)\ \text{ii, iii and the condition } \langle u_n, u_1\rangle = -\cos\theta \,\text{hold}\right\}.$$

Let $SO(3)$ act on $A_n(\theta)$ diagonally. Then for an element $(u_1, \cdots, u_n) \in A_n(\theta)$, we may normalize u_1 and u_n to be as in (1) i. Hence (2) in fact coincides with (1).

The following result is known.

Theorem 3 ([9], Theorems A and B). *(i) For $n \ge 4$, we have $C_n(\theta) \ne \varnothing$ if and only if θ belongs to the following interval:*

$$\begin{cases} \left[2\arcsin\dfrac{1}{n-1}, \dfrac{n-2}{n}\pi\right], & \text{if } n \text{ is odd}, \\ \left[0, \dfrac{n-2}{n}\pi\right], & \text{if } n \text{ is even}. \end{cases} \tag{3}$$

(ii). Let a be an endpoint of the intervals in (3). *Then we have $C_n(a) = \{\text{one point}\}$.*

Example 4. For $n = 4$ or 5, the following results hold, where we omit the cases which can be read from Theorem 3.

i. For $0 < \theta < \frac{\pi}{2}$, we have $C_4(\theta) = \{\text{two points}\}$.

ii. The topological type of $C_5(\theta)$ is given by the following **Table 2**.

Here we define η_1 and η_2 to be the following **Figures 3** and **4**, respectively.

The proof of the example will be given at the end of §5.

In [8], the following proposition is proved using the implicit function theorem.

Proposition 5 ([8], Proposition 1). *There exists θ_0 such that for all $\theta \in \left(\theta_0, \frac{n-2}{n}\pi\right)$, the system of equations defined by* (1) i, ii *and* iii *intersect transversely. Hence for such θ, $C_n(\theta)$ carries a natural differential structure.*

Range of θ	**$2\arcsin\frac{1}{4} < \theta < \frac{\pi}{5}$**	**$\frac{\pi}{5}$**	**$\frac{\pi}{5} < \theta < \frac{\pi}{3}$**	**$\frac{\pi}{3}$**	**$\frac{\pi}{3} < \theta < \frac{3}{5}\pi$**
Topological type of $C_5(\theta)$	S^1	η_1	$S^1 \coprod S^1$	η_2	S^1

Table 2.
The topological type of $C_5(\theta)$.

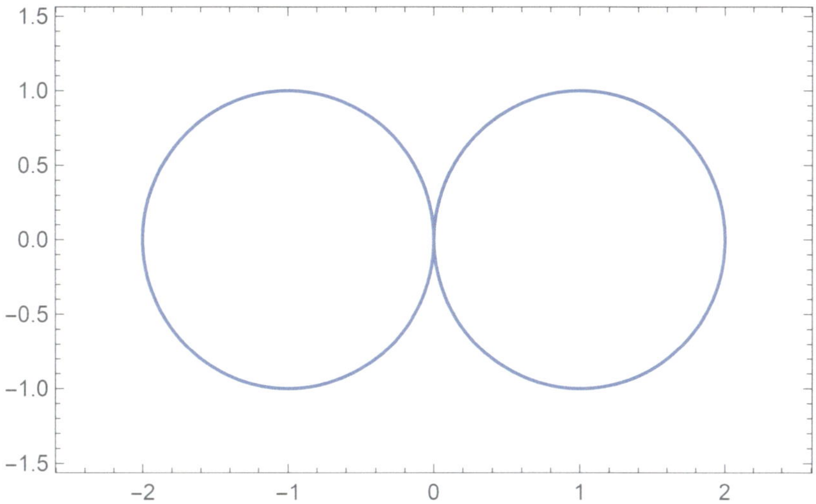

Figure 3.
The space η_1.

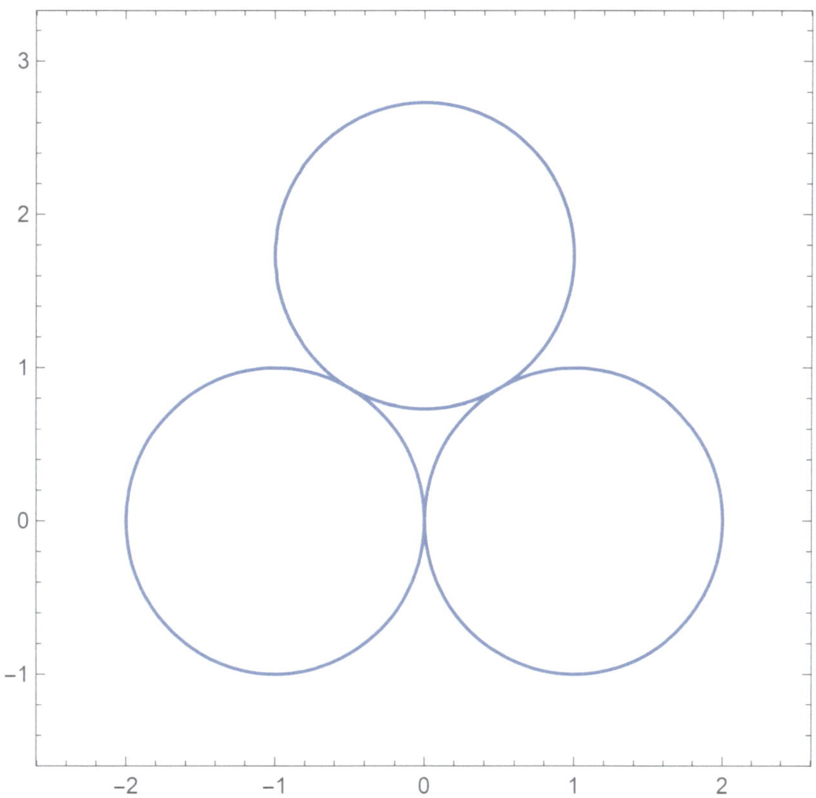

Figure 4.
The space η_2.

The main result in [8] is the following:

Theorem 6 ([8], Theorem 1)**.** *Let θ_0 be as in Proposition 5. Then for all $\theta \in \left(\theta_0, \frac{n-2}{n}\pi\right)$, $C_n(\theta)$ is homeomorphic to S^{n-4}.*

Remark 7. In [8], Theorem 6 is proved by the following method: We construct a function $f : C_n(\theta) \to \mathbb{R}$ and show that f has exactly two critical points. Then Reeb's theorem implies Theorem 6. Note that with this method, we cannot improve the assertion from homeomorphism to diffeomorphism. (See ([10], p. 25) for Reeb's theorem and remarks about it.)

The following theorem is the answer to Problem 1 (i).

Theorem A. *We equip S^{n-4} with the standard differential structure. Let θ_0 be as in Proposition 5. Then for all $\theta \in \left(\theta_0, \frac{n-2}{n}\pi\right)$, $C_n(\theta)$ is diffeomorphic to S^{n-4}.*

Next we consider Problem 1 (ii). We set

$$\alpha_n := \inf \left\{ \theta_0 \in (0,\pi)\ \mathrm{C_n}(\theta) \cong \mathrm{S}^{n-4} \text{ holds for all } \theta \in \left(\theta_0, \frac{\mathrm{n}-2}{\mathrm{n}}\pi\right) \right\}.$$

Here in what follows, the notation $X \cong Y$ means that X is homeomorphic to Y. Note that among the values of θ_0 in Theorem 6, α_n is the best possible one.

The following result is known.

Theorem 8 ([11]). *(i) We have $\alpha_n = \frac{n-4}{n-2}\pi$ for $4 \le n \le 7$.*

(ii). We have $\alpha_8 < \frac{5}{7}\pi$.

Remark 9. About Theorem 8 (i), we can read α_4 and α_5 from the above Example 4, and α_6 from **Table 4** in Theorem D below.

From Theorem 8, we naturally encounter the following:

Question 10. (i) Is it true that $\alpha_n = \frac{n-4}{n-2}\pi$ holds for $n \ge 4$?

(ii) Is it true that $\alpha_n < \frac{n-3}{n-1}\pi$ holds for $n \ge 4$? Note that if (i) is true then (ii) holds automatically.

The following theorem is the answer to Problem 1 (ii).

Theorem B. *For $4 \le n \le 14$, the following* **Table 3** *holds.*

The following theorem is the answer to Question 10.

Theorem C. *(i) The statement in Question 10* (i) *is false for $n \ge 8$. In fact, we have $\frac{n-4}{n-2}\pi < \alpha_n$ for $n \ge 8$.*

(ii) The statement in Question 10 (ii) *is false for $n \ge 13$. In fact, we have $\frac{n-3}{n-1}\pi < \alpha_n$ for $n \ge 13$.*

The following theorem is the answer to Problem 1 (iii).

Theorem D. *The topological type of $C_6(\theta)$ is given by the following* **Table 4**, *where we omit the cases which can be read from Theorem 3.*

n	α_n	$\frac{n-4}{n-2}\pi$	$\frac{n-3}{n-1}\pi$	$\frac{n-2}{n}\pi$
4	0	0	0.333π	0.500π
5	0.333π	0.333π	0.500π	0.600π
6	0.500π	0.500π	0.600π	0.667π
7	0.600π	0.600π	0.667π	0.714π
8	0.676π	0.667π	0.714π	0.750π
9	0.729π	0.714π	0.750π	0.778π
10	0.767π	0.750π	0.778π	0.800π
11	0.795π	0.778π	0.800π	0.818π
12	0.817π	0.800π	0.818π	0.833π
13	0.834π	0.818π	0.833π	0.846π
14	0.848π	0.833π	0.846π	0.857π

Table 3.
The value of α_n for $4 \le n \le 14$.

Range of θ	$0<\theta<\frac{\pi}{3}$	$\frac{\pi}{3}<\theta<\frac{\pi}{2}$	$\frac{\pi}{2}<\theta<\frac{2}{3}\pi$
Topological type of $C_6(\theta)$	$\underset{3}{\#}(S^1\times S^1)$	$\underset{3}{\#}(S^1\times S^1)$	S^2

Table 4.
The topological type of $C_6(\theta)$.

Remark 11. (i) As indicated in Problem 1 (iii), not only S^2 but also $\underset{3}{\#}(S^1\times S^1)$ appears in **Table 4** as a topological type of $C_6(\theta)$

(ii) We can also determine the topological type of $C_6\left(\frac{\pi}{2}\right)$ and $C_6\left(\frac{\pi}{3}\right)$. (See Remark 18 in §5.) In particular, they have singular points.

3. Proof of Theorem A

Following the method of [12], we set

$$X_n := \left\{(P,\theta)\in\left(S^2\right)^n\times\left(0,\frac{n-2}{n}\pi\right] \mid P\in C_n(\theta)\right\}. \tag{4}$$

We define the function $\mu : X_n \to \mathbb{R}$ by

$$\mu(P,\theta)=\theta. \tag{5}$$

Note that for all $\theta\in(0,\pi]$, we have

$$\mu^{-1}(\theta)=C_n(\theta). \tag{6}$$

The following proposition holds:

Proposition 12. (i) *Let θ_0 be as in Proposition 5. Then the space $\mu^{-1}\left(\theta_0,\frac{n-2}{n}\pi\right]$ is a manifold, where $\mu^{-1}\left(\theta_0,\frac{n-2}{n}\pi\right]$ denotes the inverse image of the interval.*

(ii) *Any element of $\mu^{-1}\left(\theta_0,\frac{n-2}{n}\pi\right)$ is a regular point of μ.*

(iii) *Consider the case $n=8$. Then $C_8\left(\frac{6}{8}\pi\right)$ is a non-degenerate critical point of μ.*

In order to prove the proposition, we need a lemma. We set

$$D_n := \left\{(u_1,\cdots,u_{n-2},\theta)\in\left(S^2\right)^{n-2}\times(0,\pi) \mid \text{the following i and ii hold}\right\}.$$

i. $u_1=(1,0,0)$.

ii. $\langle u_i,u_{i+1}\rangle=-\cos\theta$ for $1\le i\le n-3$.

Lemma 13. *(i) There is a diffeomorphism*

$$f:\left(S^1\right)^{n-3}\times(0,\pi)\underset{\cong}{\to}D_n.$$

(ii) We define the map $L : D_n \to \mathbb{R}$ by

$$L(u_1,\cdots,u_{n-2},\theta)=\|(-\cos\theta,-\sin\theta,0)+\sum_{i=1}^{n-2}u_i\|^2.$$

Then we have the following commutative diagram:

$$\begin{array}{ccc} X_n & \xrightarrow[\cong]{g} & (L\circ f)^{-1}(1) \\ & {}_{\mu}\searrow \quad \swarrow_{p} & \\ & \mathbb{R} & \end{array} \tag{7}$$

Here the maps p and g are defined as follows.

- *Let*

$$Pr : \left(S^1\right)^{n-3} \times (0,\pi) \to \mathbb{R} \tag{8}$$

be the projection to the $(0,\pi)$*-component and we denote by p the restriction of Pr to* $(L\circ f)^{-1}(1)$:

$$p := Pr|_{(L\circ f)^{-1}(1)}.$$

- *The map g is a homeomorphism which will be defined in* (13).

(iii) For all $\theta\in(0,\pi)$*, the restriction of the map g in* (7) *naturally induces a homeomorphism*

$$g|_{C_n(\theta)} : C_n(\theta) \underset{\cong}{\to} p^{-1}(\theta). \tag{9}$$

Proof of Lemma 13*:* (i) From an element

$$\left(e^{i\phi_1}, \cdots, e^{i\phi_{n-3}}, \theta\right) \in \left(S^1\right)^{n-3} \times (0,\pi),$$

we construct the element $(u_1, \cdots, u_{n-2}, \theta) \in D_n$ as follows: In the process of constructing u_i, we also construct the elements $v_i \in S^2$ such that $\langle u_i, v_i\rangle = 0$. We set

$$u_{i+1} := -(\cos\theta)u_i + (\sin\theta\cos\phi_i)v_i + (\sin\theta\sin\phi_i)u_i \times v_i \tag{10}$$

and

$$v_{i+1} := -(\sin\theta)u_i - (\cos\theta\cos\phi_i)v_i - (\cos\theta\sin\phi_i)u_i \times v_i, \tag{11}$$

where $u_i \times v_i$ denotes the cross product.

In (10) and (11) for $i = 1$, we set $u_1 := (1,0,0)$ and $v_1 := (0,1,0)$. Then we obtain u_2 and v_2. Next using (10) and (11) for $i = 2$, we obtain u_3 and v_3. Repeating this process, we obtain u_i and v_i for $1 \le i \le n-2$. Now we define f by

$$f\left(e^{i\phi_1}, \cdots, e^{i\phi_{n-3}}, \theta\right) := (u_1, \cdots, u_{n-2}, \theta).$$

From the construction, f is a diffeomorphism.

(ii) We define the map $h : X_n \to L^{-1}(1)$ by

$$h(u_1, \cdots, u_n, \theta) := (u_1, \cdots, u_{n-2}, \theta). \tag{12}$$

Since $(u_1, \cdots, u_n)$ is an element of $C_n(\theta)$, the right-hand side of (12) is certainly an element of $L^{-1}(1)$. It is clear that h is a homeomorphism. Hence if we define the map g by

$$g := f^{-1} \circ h, \tag{13}$$

then g is also a homeomorphism. From the construction, it is clear that the diagram (7) is commutative.

(iii) The item is clear from (6) and the diagram (7).

Proof of Proposition 12: Recall that $(L \circ f)^{-1}(1)$ in (7) is a subspace of $(S^1)^{n-3} \times (0, \pi)$. In order to prove Proposition 12, we calculate in the universal covering space. Let

$$q : \mathbb{R}^{n-3} \times (0, \pi) \to (S^1)^{n-3} \times (0, \pi)$$

be the universal covering space and we define the map

$$\widetilde{Pr} : \mathbb{R}^{n-3} \times (0, \pi) \to \mathbb{R}$$

by $\widetilde{Pr} := Pr \circ q$, where the map Pr is defined in (8). Then in addition to (7), we have the following commutative diagram:

$$\begin{array}{ccccccc} \mathbb{R}^{n-3} \times (0, \pi) & \xrightarrow{q} & (S^1)^{n-3} \times (0, \pi) & \xrightarrow[\cong]{f} & D_n & \xrightarrow{L} & \mathbb{R} \\ & \searrow^{\widetilde{Pr}} & \swarrow_{Pr} & & & & \\ & \mathbb{R} & & & & & \end{array} \tag{14}$$

(i) Let $x \in \mathbb{R}^{n-3} \times (0, \pi)$ be any element which satisfies the condition

$$q(x) \in p^{-1}\left(\theta_0, \frac{n-2}{n}\pi\right]. \tag{15}$$

Note that if we use the diagram (14), then (15) is equivalent to saying that

$$\widetilde{Pr}(x) \in \left(\theta_0, \frac{n-2}{n}\pi\right] \quad \text{and} \quad (L \circ f \circ q)(x) = 1.$$

We set

$$\text{grad}_x(L \circ f \circ q) := \left(\frac{\partial(L \circ f \circ q)}{\partial \phi_1}(x), \cdots, \frac{\partial(L \circ f \circ q)}{\partial \phi_{n-3}}(x), \frac{\partial(L \circ f \circ q)}{\partial \theta}(x)\right). \tag{16}$$

In order to prove Proposition 12 (i), it will suffice to prove that

$$\text{grad}_x(L \circ f \circ q) \neq (0, \cdots, 0) \tag{17}$$

(a) *The case when* $\widetilde{Pr}(x) = \frac{n-2}{n}\pi$.

We claim that x has the form

$$x = \left(0, \cdots, 0, \frac{n-2}{n}\pi\right). \tag{18}$$

To prove this, recall the homeomorphism $g|_{C_n(\theta)}$ was defined in (9). Since $\left(g|_{C_n(\theta)}\right)^{-1}(q(x))$ is the regular n-gon, (18) follows.

We shall prove that

$$\mathrm{grad}_x(L\circ f\circ q) = (0,\cdots,0,r) \tag{19}$$

for some positive real number r.

First, note that the real-valued function $(L\circ f\circ q)\left(\phi_1,\cdots,\phi_{n-3},\frac{n-2}{n}\pi\right)$ takes the minimum value 0 at $(\phi_1,\cdots,\phi_{n-3}) = (0,\cdots,0)$. Hence the first $(n-3)$-terms of the both sides of (19) coincide.

Second, direct computations show that

$$(L\circ f\circ q)(0,\cdots,0,\theta) = \begin{cases} 4\left(\displaystyle\sum_{i=1}^{m}(-1)^i \sin\frac{2i-1}{2}\theta\right)^2, & \text{if } n = 2m+1, \\ \left(1+2\displaystyle\sum_{i=1}^{m-1}(-1)^i \cos i\theta\right)^2, & \text{if } n = 2m. \end{cases} \tag{20}$$

The number r in (19) equals to the derivative of (20) at $\theta = \frac{n-2}{n}\pi$. It is easy to see that

$$\begin{cases} (-1)^i \cos\dfrac{(2i-1)(2m-1)}{4m+2}\pi < 0, & \text{for } 1\le i\le m, \\ (-1)^{i+1}\sin\dfrac{i(2m-2)}{2m}\pi > 0, & \text{for } 1\le i\le m-1 \end{cases} \tag{21}$$

and

$$\begin{cases} \displaystyle\sum_{i=1}^{m}(-1)^i \sin\frac{(2i-1)(2m-1)}{4m+2}\pi = -\frac{1}{2}, \\ 1+2\displaystyle\sum_{i=1}^{m-1}(-1)^i \cos\frac{i(2m-2)}{2m}\pi = 1. \end{cases} \tag{22}$$

Using (21) and (22), we can check that the derivative of (20) at $\theta = \frac{n-2}{n}\pi$ is positive, i.e., r is positive. Thus we have obtained (19). This completes the proof of (17) for the case (a).

(b) *The case when* $\widetilde{Pr}(x) \in \left(\theta_0, \frac{n-2}{n}\pi\right)$.

By Proposition 5, we have

$$\left(\frac{\partial(L\circ f\circ q)}{\partial\phi_1}(x),\cdots,\frac{\partial(L\circ f\circ q)}{\partial\phi_{n-3}}(x)\right) \neq (0,\cdots,0). \tag{23}$$

Then using (16), we obtain (17). This completes the proof of (17) for the case (b), and hence also that of (i).

(ii) In order to prove by contradiction, assume that $\mu^{-1}\left(\theta_0,\frac{n-2}{n}\pi\right)$ contains a critical point of μ. Then using (7), $p^{-1}\left(\theta_0,\frac{n-2}{n}\pi\right)$ contains a critical point of p. Lifting to the universal covering space using (14), there exists an element $x\in\mathbb{R}^{n-3}\times(0,\pi)$ which satisfies the following two items:

- We have $\widetilde{Pr}(x) \in \left(\theta_0,\frac{n-2}{n}\pi\right)$.

- The point x is a critical point of the function $\widetilde{Pr}$ under the constraint

$$(L \circ f \circ q)(\phi_1, \cdots, \phi_{n-3}, \theta) = 1. \tag{24}$$

We apply the Lagrange multiplier method to (24). Since $\widetilde{Pr}(\phi_1, \cdots, \phi_{n-3}, \theta) = \theta$, there exists $\lambda \in \mathbb{R}$ such that

$$(0, \cdots, 0, 1) = \lambda \operatorname{grad}_x(L \circ f \circ q). \tag{25}$$

We compare the first $(n-3)$-components of the both sides of (25). Then by (23), we have $\lambda = 0$. But this contradicts the last component of (25). Hence (25) cannot occur. This completes the proof of (ii).

(iii) Consider the Eq. (24). Using the implicit function theorem, we may assume that θ is a function with variables ϕ_i $(1 \le i \le n-3)$: $\theta = \theta(\phi_1, \cdots, \phi_{n-3})$. Note that

$$\widetilde{Pr}(\phi_1, \cdots, \phi_{n-3}, \theta(\phi_1, \cdots, \phi_{n-3})) = \theta(\phi_1, \cdots, \phi_{n-3}).$$

Hence it will suffice to prove the following result for $n = 8$:

$$\left| \left(\frac{\partial^2 \theta(\phi_1, \cdots, \phi_{n-3})}{\partial \phi_i \partial \phi_j}(0, \cdots, 0) \right)_{1 \le i,j \le n-3} \right| \neq 0, \tag{26}$$

where $||$ denotes the determinant. Computing by the method of second implicit derivative, we see that the value of the left-hand side of (26) for $n = 8$ is $\frac{1-\sqrt{2}}{16384}$. Hence (26) holds for $n = 8$. This completes the proof of (iii), and hence also that of Proposition 12.

In order to prove Theorem A, we recall the following:

Theorem 14 ([13], Corollary B). *For $d \ge 2$, let M be a d-dimensional smooth manifold without boundary and $F : M \to \mathbb{R}$ a smooth function. We set $\max F(M) = m$ and assume that m is attained by unique point $z \in M$. Let $a \in \mathbb{R}$ satisfy the following four conditions:*

i. $a < m$.

ii. $F^{-1}[a, m]$ is compact.

iii. *There are no critical points in $F^{-1}[a, m)$.*

iv. $d \neq 5$.

Then there is a diffeomorphism $F^{-1}(a) \cong S^{d-1}$.

Remark 15. For the proof of Theorem 14, the h-cobordism theorem (see [14], p. 108, Proposition A) is crucial. Hence we cannot drop the condition $d \neq 5$.

Proof of Theorem A: First, we consider the case $n \neq 8$.

- For F in Theorem 14, we consider

$$\mu : \mu^{-1}\left(\theta_0, \frac{n-2}{n}\pi\right] \to \mathbb{R}.$$

More precisely, we denote the restriction of μ in (5) to $\mu^{-1}\left(\theta_0, \frac{n-2}{n}\pi\right]$ by the same symbol μ. Note that from the definition of F, z in Theorem 14 is $C_n\left(\frac{n-2}{n}\pi\right)$.

- By Proposition 12 (i), $\mu^{-1}\left(\theta_0, \frac{n-2}{n}\pi\right]$ is in fact a manifold.
- For a in Theorem 14, we consider any element θ in $\left(\theta_0, \frac{n-2}{n}\pi\right)$.

Below we check that the conditions i, ii, iii and iv in Theorem 14 are satisfied. The items i and ii are clear. The item iii follows from Proposition 12 ii. The item iv follows from the following argument: Since $\dim X_n = n - 3$, we have $\dim \mu^{-1}\left(\theta_0, \frac{n-2}{n}\pi\right] \neq 5$ if and only if $n \neq 8$.

Now we can apply Theorem 14 and obtain that $\mu^{-1}(\theta)$ is diffeomorphic to S^{n-4} if θ satisfies that $\theta_0 < \theta < \frac{n-2}{n}\pi$. By (6), this is equivalent to saying that $C_n(\theta)$ is also diffeomorphic to S^{n-4}. This completes the proof of Theorem A for $n \neq 8$.

Second, we consider the case $n = 8$. If we apply the Morse lemma to Proposition 12 (iii), then we obtain that $\mu^{-1}(\theta)$ is diffeomorphic to S^4 if θ satisfies that $\theta_0 < \theta < \frac{6}{8}\pi$. Hence $C_8(\theta)$ is also diffeomorphic to S^4. This completes the proof of Theorem A for $n = 8$.

4. Proofs of Theorems B and C

Proof of Theorem B: In ([8], Lemma 1), certain conditions on an element $(u_1, \cdots, u_n)$ of $C_n(\theta)$ are listed. For example, a condition is given by

$$u_2 = u_n. \tag{27}$$

Let Λ be the set of the conditions. For $\lambda \in \Lambda$, we set

$$\Theta_\lambda := \inf \ \left\{\theta_0 \in \left(0, \frac{n-2}{n}\pi\right) \ \middle| \ \text{for all } \theta \in \left(\theta_0, \frac{\mathrm{n}-2}{\mathrm{n}}\pi\right), \mathrm{C_n}(\theta) \text{ does not} \right. \\ \left. \text{containan element which satisfies the condition } \lambda\right\}. \tag{28}$$

Using this, we set

$$\beta_n := \max \ \{\Theta_\lambda \ | \ \lambda \in \Lambda\}. \tag{29}$$

Then it is proved in ([8], Proposition 1) that

$$\alpha_n = \beta_n. \tag{30}$$

We explain how to compute Θ_λ. As an example of λ, we consider the condition (27). We construct the continuous function

$$R_\lambda : (0, \pi) \to \mathbb{R} \tag{31}$$

which satisfies the following two properties:

a. We have $R_\lambda(\theta) \geq 0$ for all θ.

b. An element $\theta \in (0, \pi)$ satisfies $R_\lambda(\theta) = 0$ if and only if $C_n(\theta)$ contains an element which satisfies the condition (27).

In order to construct R_λ in (31), we first fix θ and define the space $Y_n(\theta)$ as follows:

$$Y_n(\theta) := \{(u_1, \cdots, u_n) \in (S^2)^n \mid \text{the following i and ii hold}\}.$$

i. $u_1 = (1, 0, 0)$ and $u_2 = u_n = (-\cos\theta, -\sin\theta, 0)$.

ii. $\langle u_i, u_{i+1} \rangle = -\cos\theta$ for $2 \le i \le n-3$.

Second, we define the function $r_\lambda : Y_n(\theta) \to \mathbb{R}$ as follows: For $(u_1, \cdots, u_n) \in Y_n(\theta)$, we set

$$r_\lambda(u_1, \cdots, u_n) := \left\| \sum_{i=1}^{n} u_i \right\|. \tag{32}$$

Third, we define R_λ in (31) by

$$R_\lambda(\theta) := \min r_\lambda(Y_n(\theta)).$$

Below we check the above properties a and b of R_λ.
The item a is clear.
In order to prove the item b, we claim the following identification holds:

$$r_\lambda^{-1}(0) = \{(u_1, \cdots, u_n) \in C_n(\theta) \mid u_2 = u_n\}. \tag{33}$$

In fact, an element $(u_1, \cdots, u_n) \in Y_n(\theta)$ belongs to $r_\lambda^{-1}(0)$ if and only if (1) ii holds. Hence (33) follows.

Now the item b is clear from (33). Thus we have checked the above properties a and b.

Next using the properties a and b, we can describe Θ_λ in (28) as

$$\Theta_\lambda = \max\{\theta \in (0, \pi) \mid R_\lambda(\theta) = 0\}. \tag{34}$$

From the constructions in (10) and (11), we have

$$Y_n(\theta) \cong (S^1)^{n-4} \times S^2.$$

Using this fact, we can compute the right-hand side of (34) for $n \le 14$.

By a similar method, we compute Θ_λ for each $\lambda \in \Lambda$. Then from the definition of β_n in (29), we can determine β_n. Finally, using (30), we obtain α_n. This completes the proof of Theorem B.

Remark 16. In the above proof of Theorem B, the identification (33) is crucial. Although $r_\lambda^{-1}(0)$ is a critical submanifold of the function r_λ in (32), this fact allows us to compute the right-hand side of (34) for $n \le 14$. See §6 (ii) for further remarks.

Proof of Theorem C: The theorem is clear from **Table 3**.

5. Proof of Theorem D

The following proposition is a refinement of Proposition 12 for $n = 6$.

Proposition 17. *(i) The space X_6 is a manifold, where X_n is defined in* (4).

(ii) The interior angle θ is a critical point of μ if and only if θ equals to $\frac{\pi}{3}$, $\frac{\pi}{2}$ or $\frac{2}{3}\pi$.

Proof: We can prove the proposition is the same way as in Proposition 12. Since the dimension is low, we can perform direct computations.

We apply the first fundamental theorem of Morse theory to Proposition 17. Then we obtain the following assertion: If θ_1 and θ_2 belong to the same interval from the three intervals $\left(0, \frac{\pi}{3}\right)$, $\left(\frac{\pi}{3}, \frac{\pi}{2}\right)$ and $\left(\frac{\pi}{2}, \frac{2}{3}\pi\right)$, then we have $C_6(\theta_1) \cong C_6(\theta_2)$.

The homeomorphism (9) tells us that in order to determine the topological type of $C_6(\theta)$, it will suffice to determine the topological type of $p^{-1}(\theta)$ for $n = 6$. For a fixed $\psi \in [0, 2\pi]$, we set

$$M_\theta(\psi) := \left\{ \left(e^{i\phi_1}, e^{i\phi_2}, e^{i\phi_3}, \theta\right) \in p^{-1}(\theta) \;\middle|\; \phi_1 = \psi \right\}.$$

Since $M_\theta(\psi)$ is a one dimensional object, it is not to difficult to draw its figure. The results are given as follows.

(i) *The case when* $\frac{\pi}{2} < \theta < \frac{2}{3}\pi$.

There exists ω in $(0, \pi)$ such that the following homeomorphism holds:

$$M_\theta(\psi) \cong \begin{cases} \{\text{one point}\}, & \text{if } \psi = \omega \text{ or } 2\pi - \omega, \\ \varnothing, & \text{if } \omega < \psi < 2\pi - \omega, \\ S^1, & \text{otherwise.} \end{cases}$$

From this, we have $p^{-1}(\theta) \cong S^2$. The figure of $C_6(\theta)$ is given by the following **Figure 6**.

(ii) *The case when* $\frac{\pi}{3} < \theta < \frac{\pi}{2}$.

There exists ω in $(0, \pi)$ such that the following homeomorphism holds:

$$M_\theta(\psi) \cong \begin{cases} \sigma, & \text{if } \psi = \omega \text{ or } 2\pi - \omega. \\ S^1, & \text{otherwise,} \end{cases} \tag{35}$$

Here we set

$$\sigma := \bigcup_{i=1}^{2} \left\{ (x, y) \in \mathbb{R}^2 \;\middle|\; x^2 + i^2 y^2 = 1 \right\}.$$

(The figure of σ is given by the following **Figure 5(b)**.)

We claim that the four intersection points in $M_\theta(\omega) \cup M_\theta(2\pi - \omega)$ are saddle points of $p^{-1}(\theta)$. In fact, for a sufficiently small positive real number ε, the following **Figure 5 (a)–(c)** give the shape of $M_\theta(\omega - \varepsilon)$, $M_\theta(\omega)$ and $M_\theta(\omega + \varepsilon)$, respectively. (The deformation of the shape of $M_\theta(\psi)$ when ψ is near $2\pi - \omega$ is also given by **Figure 5**.) Now from **Figure 5**, we see that the four intersection points are in fact saddle points.

Since we identify $M_\theta(0)$ with $M_\theta(2\pi)$, (35) and **Figure 5** give the homeomorphism $p^{-1}(\theta) \cong \underset{3}{\#}\left(S^1 \times S^1\right)$. The figure of $C_6(\theta)$ is given by the following **Figure 7**.

(iii) *The case when* $0 < \theta < \frac{\pi}{3}$.

The topological type of $M_\theta(\psi)$ is the same as (35). Hence the argument in (ii) remains valid.

Remark 18. We determine the topological type of $C_6\left(\frac{\pi}{2}\right)$ and $C_6\left(\frac{\pi}{3}\right)$.

(i) The figure of $C_6\left(\frac{\pi}{2}\right)$ is given by the following **Figure 8**.

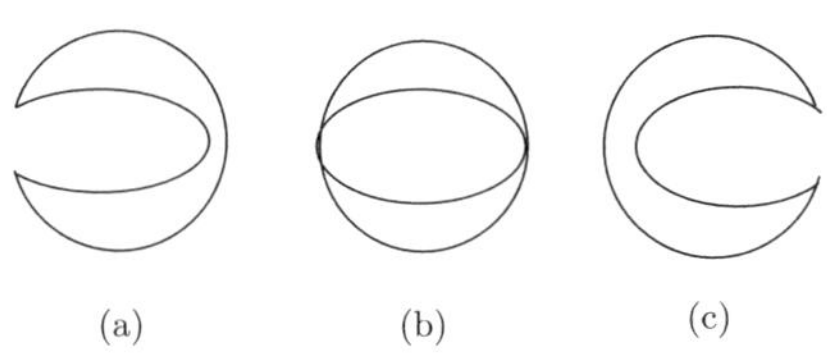

Figure 5.
(a) $M_\theta(\omega - \varepsilon)$*; (b)* $M_\theta(\omega)$*; (c)* $M_\theta(\omega + \varepsilon)$.

DOI: http://dx.doi.org/10.5772/intechopen.100723

Thus $C_6\left(\frac{\pi}{2}\right)$ is homeomorphic to the orbit space $S^2/\sim$, where the equivalence relation is generated by

$$(-1,0,0) \sim (1,0,0), \quad (0,-1,0) \sim (0,1,0) \quad \text{and} \quad (0,0,-1) \sim (0,0,1).$$

In particular, $C_6\left(\frac{\pi}{2}\right)$ has three singular pints.

As θ approaches $\frac{\pi}{2}$ from below, each center of the three handles in **Figure 7** shrinks. And when $\theta = \frac{\pi}{2}$, each center pinches to a point and we obtain **Figure 8**. If θ increases further from $\frac{\pi}{2}$, then the pinched point separates and we obtain **Figure 6**.

(ii) The figure of $C_6\left(\frac{\pi}{3}\right)$ is given by the following **Figure 9**.

The space $C_6\left(\frac{\pi}{3}\right)$ contains subspaces

$$N_1, \ N_2 \ \text{and} \ N_3 \tag{36}$$

which satisfy the following three properties:

- $N_1 \cong S^1 \times S^1, .N_2 \cong S^1 \times S^1$ and $N_3 \cong \underset{2}{\#}\left(S^1 \times S^1\right)$
- $\bigcup_{i=1}^{3} N_i = C_6\left(\frac{\pi}{3}\right)$.
- $\bigcup_{i<j} \left(N_i \cap N_j\right) \cong \bigcup_{i=1}^{3} \left\{(x,y) \in \mathbb{R}^2 \mid x^2 + i^2 y^2 = 1\right\}$.

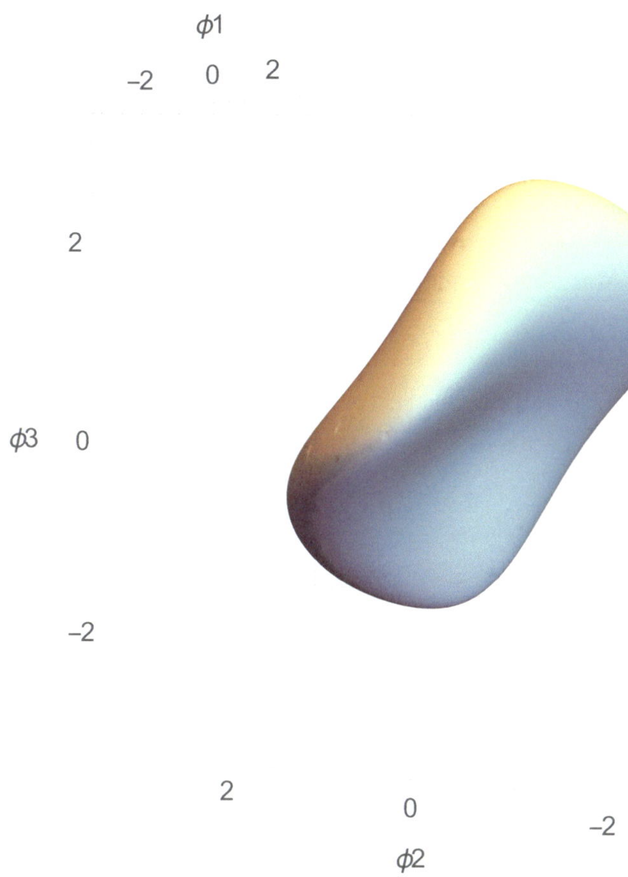

Figure 6.
The space $C_6(\theta)$ for $\frac{\pi}{2} < \theta < \frac{2}{3}\pi$.

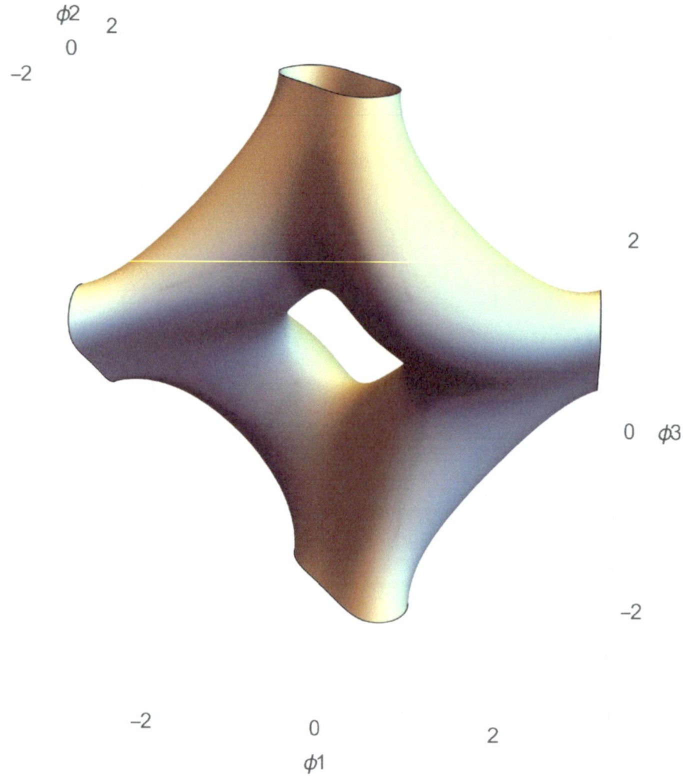

Figure 7.
The space $C_6(\theta)$ for $\frac{\pi}{3} < \theta < \frac{\pi}{2}$, where we identify the opposite boundaries.

Figure 8.
The space $C_6\left(\frac{\pi}{2}\right)$, where we identify the opposite vertices.

DOI: http://dx.doi.org/10.5772/intechopen.100723

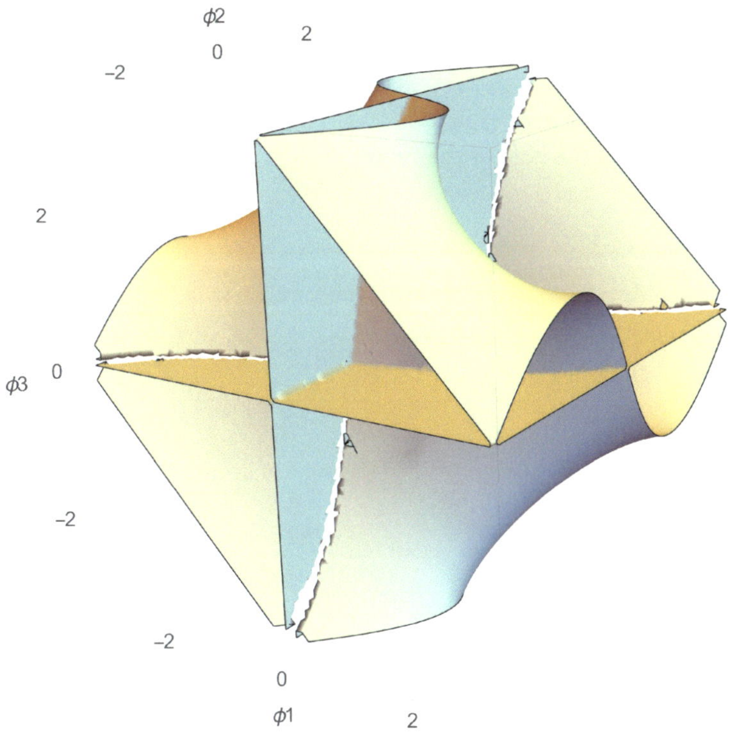

Figure 9.
The space $C_6\left(\frac{\pi}{3}\right)$.

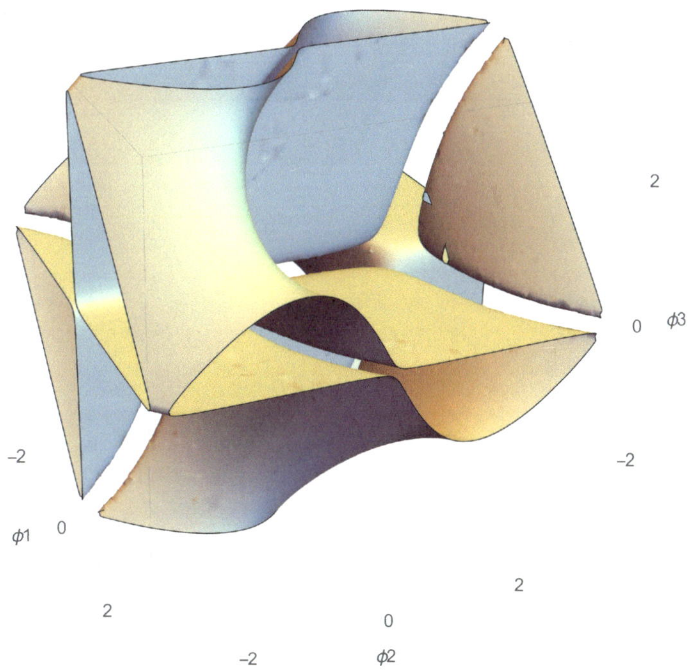

Figure 10.
The space $C_6\left(\frac{\pi}{3} \pm \varepsilon\right)$.

The figure of $C_6\left(\frac{\pi}{3} \pm \varepsilon\right)$ is given by **Figure 10** above.

As θ approaches $\frac{\pi}{3}$, a cross-section of the four tubes in **Figure** 7 becomes a union of two circles: In the notation of (36), one circle becomes a handle of N_3. And the other circle is a subspace of $N_1 \cup N_2$.

On the other hand, the hole of the center of **Figure** 7 becomes a subspace of $N_1 \cup N_2$.

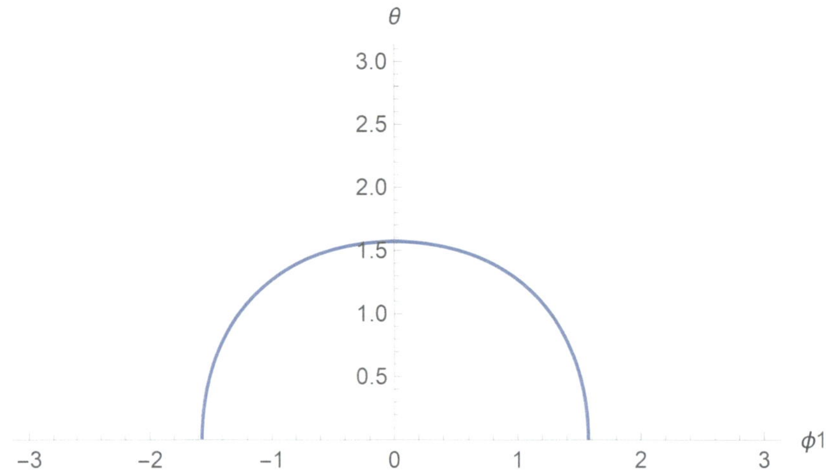

Figure 11.
The space X_4.

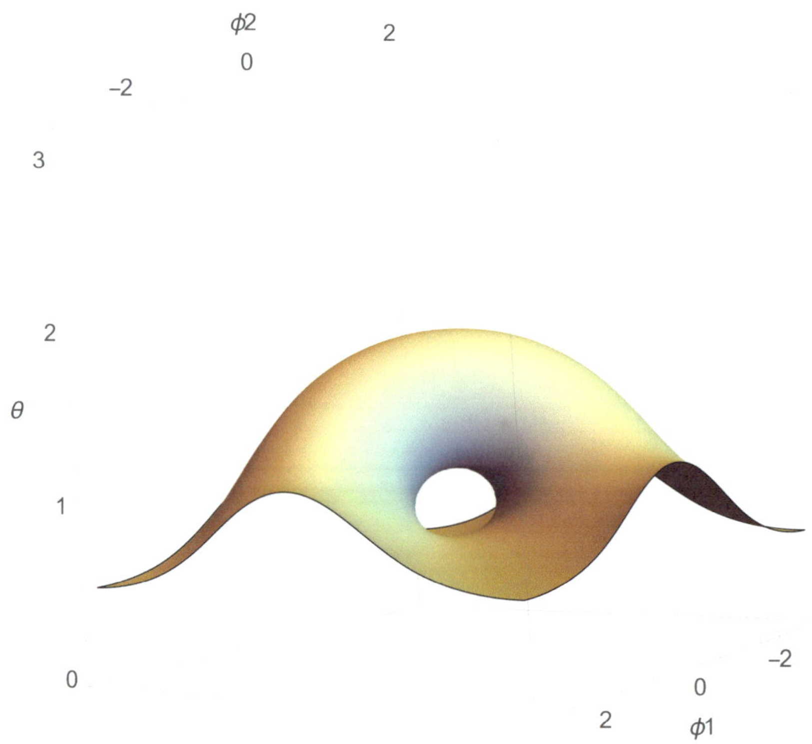

Figure 12.
The space X_5.

Proof of Example 4: We can prove the example in the same way as in Theorem D. We can also prove by the following method. Recall that the space X_n was defined in (4). The figures of X_4 and X_5 are given by **Figures 11** and **12** above, respectively.

The identification (6) tells us that each level set of **Figure 11** gives $C_4(\theta)$, and that of **Figure 12** gives $C_5(\theta)$. Thus we obtain Example 4.

6. Conclusions

i. We have the following comments about the proof of Theorem A. Recall that for the proof of Theorem A given in §3, we used Proposition 5 but we did not use Theorem 6. In other words, we did not use Reeb's theorem. Instead, we used Theorem 14, for which the h-cobordism theorem is crucial. From the computations for small n, it seems that (26) holds for all n. If we could prove this, then we obtain a proof which uses only the Morse lemma. We pose the following question: Is it possible to prove (26) for all n?

ii. We have the following comments about the proof of Theorem B. Recalling (23) and (24), we consider the following system of equations:

$$\left(\frac{\partial(L\circ f\circ q)}{\partial\phi_1}(x), \cdots, \frac{\partial(L\circ f\circ q)}{\partial\phi_{n-3}}(x)\right) = (0, \cdots, 0) \tag{37}$$

and

$$(L\circ f\circ q)(\phi_1, \cdots, \phi_{n-3}, \theta) = 1. \tag{38}$$

If we could solve the system of Eqs. (37) and (38) with respect to the variables $\phi_1, \cdots, \phi_{n-3}$ and θ, then we could determine for which θ, $C_n(\theta)$ has a singular point and the set of singular points of $C_n(\theta)$. In particular, we obtain Proposition 5. But it is not easy to solve a system of equations even if we can use a computer. Hence, as we remarked in Remark 16, we have given the proof of Theorem B such as in §4. We pose the following question: Is it possible to solve the system of Eqs. (37) and (38) with respect to the variables $\phi_1, \cdots, \phi_{n-3}$ and θ?

Author details

Yasuhiko Kamiyama
Faculty of Science, Department of Mathematical Sciences, University of the Ryukyus, Okinawa, Japan

*Address all correspondence to: kamiyama@sci.u-ryukyu.ac.jp

References

[1] Kapovich M, Millson J. The symplectic geometry of polygons in the Euclidean space. Journal of Differential Geometry 1996; 44: 479–513. DOI: 10.4310/jdg/1214459218 https://projecteuclid.org/journals/journal-of-differential-geometry/volume-44/issue-3/The-symplectic-geometry-of-polygons-in-Euclidean-space/10.4310/jdg/1214459218.full [Accessed: 30 September 2021]

[2] Hausmann J.-C, Knutson A. The cohomology ring of polygon spaces. Annales de l'Institut Fourier 1998; 48; 281–321. DOI: 10.5802/aif.1619 https://aif.centre-mersenne.org/item/AIF_1998__48_1_281_0 [Accessed: 30 September 2021]

[3] Farber M. Invitation to Topological Robotics. Zurich Lectures in Advanced Mathematics: European Mathematical Society; 2008. 143p. DOI: 10.4171/054

[4] Crippen G. Exploring the conformation space of cycloalkanes by linearized embedding. Journal of Computational Chemistry 1992; 13: 351–361. DOI:10.1002/jcc.540130308 https://www.researchgate.net/publication/30839993_Exploring_the_conformation_space_of_cycloalkanes_by_ linearized_embedding [Accessed: 30 September 2021]

[5] O'Hara J. The configuration space of equilateral and equiangular hexagons. Osaka Journal of Mathematics 2013; 50: 477–489. DOI:10.18910/25093 https://projecteuclid.org/journals/osaka-journal-of-mathematics/ volume-50/issue-2/The-configuration-space-of-equilateral-and-equiangular- hexagons/ojm/1371833496.full [Accessed: 30 September 2021]

[6] Kamiyama Y. The configuration space of equilateral and equiangular heptagons. JP Journal of Geometry and Topology 2020, 25: 25–33 DOI:http://dx.doi.org/10.17654/GT025120025 https://www.researchgate.net/publication/347482448_THE_CONFIGURATION_SPACE_OF_EQUILATERAL_AND_EQUIANGULAR_HEPTAGONS [Accessed: 30 September 2021]

[7] Goto S, Komatsu K. The configuration space of a mathematical model for ringed hydrocarbon molecules. Hiroshima Mathematical Journal 2012; 42: 115–126. DOI: 10.32917/hmj/1333113009 https://projecteuclid.org/journals/hiroshima-mathematical-journal/volume-42/issue-1/The-configuration-space-of-a-model-for-ringed-hydrocarbon-molecules/10.32917/hmj/1333113009.full [Accessed: 30 September 2021]

[8] Goto S, Komatsu K, Yagi J. The configuration space of almost regular polygons. Hiroshima Mathematical Journal 2020; 50: 185–197. DOI: 10.32917/hmj/1595901626 https://projecteuclid.org/journals/hiroshima-mathematical-journal/ volume-50/issue-2/The-configuration-space-of-almost-regular-polygons/ 10.32917/hmj/1595901626.full [Accessed: 30 September 2021]

[9] Kamiyama Y. A filtration of the configuration space of spatial polygons. Advances and Applications in Discrete Mathematics 2019; 22: 67–74. DOI:http://dx.doi.org/10.17654/DM022010067 https://www.researchgate.net/publication/ 336117647_A_FILTRATION_OF_THE_CONFIGURATION_SPACE_OF_SPATIAL_POLYGONS [Accessed: 30 September 2021]

[10] Milnor J. Morse Theory. Annals of Mathematics Studies Volume 51: Princeton University Press; 1963. 153p. DOI: 10.1515/9781400881802

[11] Goto S, Hemmi Y, Komatsu K, Yagi J. The closed chains with spherical configuration spaces. Hiroshima Mathematical Journal 2012, 42: 253–266.

DOI: 10.32917/hmj/1345467073 https://projecteuclid.org/journals/hiroshima-mathematical-journal/ volume-42/issue-2/The-closed-chains-with-spherical-configuration-spaces/ 10.32917/hmj/1345467073.full [Accessed: 30 September 2021]

[12] Kamiyama Y. The Euler characteristic of the regular spherical polygon spaces. Homology Homotopy and Applications 2020, 22: 1–10. DOI: https://dx.doi.org/10.4310/HHA.2020.v22.n1.a1 https://www.intlpress.com/site/pub/pages/journals/items/hha/content/vols/0022/0001/a001 [Accessed: 30 September 2021]

[13] Kamiyama Y. On the level set of a function with degenerate minimum point. International Journal of Mathematics and Mathematical Sciences 2015, Article ID 493217: 6pages. DOI:10.1155/2015/493217 https://downloads.hindawi.com/journals/ijmms/2015/493217.pdf [Accessed: 30 September 2021]

[14] Milnor J. Lectures on the h-cobordism Theorem. Princeton University Press: Princeton; 1965. 122p. DOI: 10.1515/9781400878055

Chapter 5

Covers and Properties of Families of Real Functions

Lev Bukovský

Abstract

We present results on the relationships of the covering property $G(\Phi, \Psi)$ for $\Phi, \Psi \in \{\mathcal{O}, \Lambda, \Omega, \Gamma\}$ and $G \in \{S_1, S_{fin}, U_{fin}\}$ of a topological space and the selection property $G(\Phi_0, \Psi_0)$ of the corresponding family of real functions. The result already published are presented without a proof, however with a citation of the corresponding paper. We present a general Theorem that covers almost all the result of this kind. Some results about hereditary properties are enclosed. We also present Scheepers Diagram of considered covering properties for uncountable covers.

Keywords: covering properties S_1, S_{fin}, U_{fin}, selection principles S_1, S_{fin}, U^*_{fin}, Scheepers Diagram, $\mathcal{A}$-measurable function, upper $\mathcal{A}$-semimeasurable function, hereditary properties, σ-space

1. Introduction

The paper is a collection of several results concerning the equivalences of the covering properties of a topological space X and the properties of the family of real functions defined on X and related to this cover. Indeed, we shall present results about the equivalences of the covering properties $G(\Phi, \Psi)$ of a topological space X and the selection property $G(\Phi_0, \Psi_0)$ of the topological space of upper semicontinuous real functions $USC_p(X)$ on X, for $\Phi, \Psi \in \{\mathcal{O}, \Lambda, \Omega, \Gamma\}$ and $G \in \{S_1, S_{fin}, U_{fin}\}$. The upper semicontinuous functions in this connection were for first time used in [1]. In some important cases we can replace the space $USC_p(X)$ by the topological space $C_p(X)$ of continuous functions. So we obtain the equivalence of some topological property of $C_p(X)$ and a covering property of X.

It turned out that we can prove a general theorem about measurable covers in a very abstract sense that covers almost all the special results.

The covering properties $G(\Phi, \Psi)$ were essentially introduced by M. Scheepers [2] and then, for countable covers, systematically investigated by W. Just, A.W. Miller, M. Scheepers and P.J. Szeptycki [3]. Using proved equivalences they obtained the quite simple Scheepers Diagram for Countable Covers. However, not all equivalences are true for arbitrary covers, therefore the corresponding Scheepers Diagram for Arbitrary Infinite Covers, presented below, is more complicated.

Then we present some results about hereditary properties of considered covering properties for F_σ-subsets. The results are important in many considerations. Finally, inspired by the result of J. Haleš [4], we show some relationships of the hereditary property of the topological space X for any subset with the property being a σ-space.

If the presented result was already published, we present the precise citation and no proof.

2. Notations and terminology

By **a topological space** we understand an infinite Hausdorff space $\langle X, \tau \rangle$, where τ is the topology on X: the set of all open subsets of X [5]. The smallest σ-algebra containing all open sets is the family $\mathrm{BOREL}(X) = \mathrm{BOREL}$ of Borel subsets of X.

We recall some covering properties which were introduced by M. Scheepers in [2]. If $\mathcal{A}, \mathcal{B} \subseteq \mathcal{P}(Y)$ are sets of subsets of a set Y, then the *covering principle* $\mathrm{S}_1(\mathcal{A}, \mathcal{B})$ means the following: for every sequence $\langle \mathcal{U}_n : n \in \omega \rangle$ of elements of $\mathcal{A}$ and for every $n \in \omega$ there exists a $U_n \in \mathcal{U}_n$ such that $\{U_n : n \in \omega\} \in \mathcal{B}$. The covering principles $\mathrm{S}_{\mathrm{fin}}(\mathcal{A}, \mathcal{B})$ and $U_{\mathrm{fin}}(\mathcal{A}, \mathcal{B})$ are there defined in a similar way. In the case of $\mathrm{S}_{\mathrm{fin}}(\mathcal{A}, \mathcal{B})$ we choose finite sets $\mathcal{W}_n \subseteq \mathcal{U}_n$ such that $\bigcup_n \mathcal{W}_n \in \mathcal{B}$. In the case of $\mathrm{S}_{\mathrm{fin}}(\mathcal{A}, \mathcal{B})$ we chose finite sets $\mathcal{W}_n \subseteq \mathcal{U}_n$ such that $\bigcup_n \mathcal{W}_n \in \mathcal{B}$. In the case of $U_{\mathrm{fin}}(\mathcal{A}, \mathcal{B})$ we chose finite sets $\mathcal{W}_n \subseteq \mathcal{U}_n$ such that $\{\bigcup \mathcal{W}_n : n \in \omega\} \in \mathcal{B}$. Actually we tacitly assume that $Y = \mathcal{P}(X)$. If $Y \subseteq {}^X\mathbb{R}$ we define the selection property $U^*_{\mathrm{fin}}(\mathcal{A}, \mathcal{B})$ similarly, but we ask[1] that $\{\min \mathcal{W}_n : n \in \omega\} \in \mathcal{B}$.

A family $\mathcal{U}$ of subsets of X is a *cover* of X if $\bigcup \mathcal{U} = X$. A cover is *open*, if every element of the cover is an open set. A cover $\mathcal{V} \subseteq \mathcal{U}$ is said to be a *subcover* of the cover $\mathcal{U}$. If we deal with a countable cover of X we can consider it as a sequence of subsets.

If $\mathcal{A}$ is a family of subsets of X, then we denote by $\mathcal{A}_{\mathrm{ctbl}}$ the family of all countable elements of $\mathcal{A}$. If $\mathcal{A} \subseteq {}^X\mathbb{R}$ then $\mathcal{A}^+$ is the family $\{f \in \mathcal{A} : (\forall x \in X) f(x) \geq 0\}$.

We introduce four special types of covers of a given set. From some technical reasons a cover is said to be an *o-cover*. A cover $\mathcal{U}$ of a set X is a *λ-cover* if for every $x \in X$ the set $\{U \in \mathcal{U} : x \in U\}$ is infinite. A cover $\mathcal{U}$ of a set X is an *ω-cover* if $X \notin \mathcal{U}$ and for every finite $F \subseteq X$ there exists a $U \in \mathcal{U}$ such that $F \subseteq U$. Finally, a cover $\mathcal{U}$ of a set X is a *γ-cover* if $\mathcal{U}$ is infinite and for every $x \in X$ the set $\{U \in \mathcal{U} : x \notin U\}$ is finite.

We shall use the following convention. If the lower case letters φ or ψ denote one of the symbols o, λ, ω or γ, then the capital letters Φ or Ψ denote the corresponding symbol $\mathcal{O}$, Λ, Ω or Γ, respectively, and vice versa.

Assume that $\mathcal{E} \subseteq \mathcal{P}(X)$ and $\emptyset, X \in \mathcal{E}$. Dealing with the covering property U_{fin} we assume that $\mathcal{E}$ is closed under finite unions.

Let $\varphi \in \{o, \lambda, \omega, \gamma\}$. We denote by $\Phi(\mathcal{E})$ the family of all φ-covers $\mathcal{U}$ of X satisfying $\mathcal{U} \subseteq \mathcal{E}$. If $\langle X, \tau \rangle$ is an infinite Hausdorff topological space, then $\Phi(\tau)$ is simply denoted as Φ.

Evidently (we should eventually omit X from a γ-cover)

$$\Gamma(\mathcal{E}) \subseteq \Omega(\mathcal{E}) \subseteq \Lambda(\mathcal{E}) \subseteq \mathcal{O}(\mathcal{E}). \tag{1}$$

We say that a family $\mathcal{V} \subseteq \mathcal{P}(X)$ is a *refinement* of the family $\mathcal{U} \subseteq \mathcal{P}(X)$ if

$$(\forall V \in \mathcal{V})(\exists U \in \mathcal{U})\, V \subseteq U. \tag{2}$$

Let the family $\mathcal{V} \subseteq \mathcal{P}(X)$ be a refinement of the family $\mathcal{U} \subseteq \mathcal{P}(X)$. If $\mathcal{V}$ is an o- or an ω-cover, then $\mathcal{U}$ is such a cover as well. This is not true for λ- and γ-covers. If we

[1] If $\mathcal{W}$ is finite then $\min \mathcal{W} = h$, where $h(x) = \min\{f(x) : f \in \mathcal{W}\}$ for every $x \in X$. $\min \emptyset = 1$.

add finitely many subsets of X to a φ-cover, $\varphi \in \{o, \lambda, \omega, \gamma\}$, we obtain a φ-cover. Any infinite subset of a γ-cover is a γ-cover as well.

A set ${}^{\omega}\mathbb{R}$ of all sequences of reals is quasi-ordered by the **eventual dominating** relation

$$\varphi \leq^{*} \psi \equiv (\exists n_0 \in \omega)(\forall n \geq n_0)\, \varphi(n) \leq \psi(n). \tag{3}$$

A set $\mathcal{F} \subseteq {}^{\omega}\mathbb{R}$ is called **bounded** if there exists a sequence $\psi \in {}^{\omega}\mathbb{R}$ such that $\varphi \leq^{*} \psi$ for every $\varphi \in \mathcal{F}$. The set $\mathcal{F}$ is **dominating** if for every $\psi \in {}^{\omega}\mathbb{R}$ there exists a $\varphi \in \mathcal{F}$ such that $\psi \leq^{*} \varphi$. The cardinals $\mathfrak{b}$ and $\mathfrak{d}$ (see e.g. [6, 7]) are the smallest cardinalities of an unbounded and dominating family, respectively.

The set ${}^{X}\mathbb{R}$ is endowed with the product topology. For any real $a \in \mathbb{R}$ we denote by **a** the constant function defined on X with value a. There are at least two important subfamilies of ${}^{X}\mathbb{R}$: the family $C(X)$ of all continuous functions and the family $\mathrm{USC}(X)$ of all upper semicontinuous functions[2]. If they are endowed with the product topology we write $\mathrm{C}_p(X)$ and $\mathrm{USC}_p(X)$, respectively. Note that $f_n \to f$ in the product topology if and only if $(\forall x \in X) f_n(x) \to f(x)$.

We introduce three properties of an infinite family $F \subseteq {}^{X}\mathbb{R}$ of real functions.

$(\mathcal{O}_0)$ $\quad \mathbf{0}(x) = 0 \in \overline{\{f(x) : f \in F\}}$ *for every* $x \in X$.

(Ω_0) $\quad \mathbf{0} \notin F$ *and* $\mathbf{0} \in \overline{F}$ *in the prodiuct topology of* ${}^{X}\mathbb{R}$.

(Γ_0) $\quad F$ *is infinite and for every* $\varepsilon > 0$ *and for every* $x \in X$ *the set* $\{f \in F : |f(x)| \geq \varepsilon\}$ *is finite*.

Let Φ and Ψ be one of the symbols $\mathcal{O}, \Omega, \Gamma$. Similarly as for covers, we define for an infinite family $F \subseteq {}^{X}\mathbb{R}$ of real functions the set

$$\Phi_0(F) = \{H \subseteq F : H \text{ is infinite and has the property } (\Phi_0)\}. \tag{4}$$

F **satisfies the selection principle** $\mathrm{S}_1(\Phi_0, \Psi_0)$ if $\mathrm{S}_1(\Phi_0(F), \Psi_0(F))$ holds. Identifying the countable sets of functions with sequences of functions, we say that F **satisfies the sequence selection principle** $\mathrm{S}_1(\Phi_0, \Psi_0)$ if $\mathrm{S}_1((\Phi_0(F))_{\mathrm{ctbl}}, (\Psi_0(F))_{\mathrm{ctbl}})$ holds (compare, e.g., [8]: for each sequence of sequences $\left\langle \left\langle f_{n,m} : m \in \omega \right\rangle : n \in \omega \right\rangle$ of functions from F with the property (Φ_0), there exists a functions $\alpha \in {}^{\omega}\omega$ such that $\left\{ f_{n,\alpha(n)} : n \in \omega \right\}$ has the property (Ψ_0).) Similarly for $\mathrm{S}_{\mathrm{fin}}$ and U_{fin}.

3. Some results about the relationship of properties of covers and of families of real functions

The first results about covering properties and the properties of the family of continuous real functions were obtained by W. Hurewicz [9]. He proved the following two theorems.

Theorem 1 (W. Hurewicz [9]). *If X is a perfectly normal topological space then the following are equivalent.*

a. *X is a* $U_{\mathrm{fin}}(\mathcal{O}_{\mathrm{ctbl}}, \mathcal{O})$

---

[2] A function $f \in {}^{X}\mathbb{R}$ is **upper semicontinuous** if the set $\{x \in X : f(x) < a\}$ is open for every real $a \in \mathbb{R}$.

b. *For every sequence* $\langle f_n : n \in \omega \rangle$ *of continuous real functions the family* $\{\langle f_n(x) : n \in \omega \rangle : x \in X\} \subseteq {}^{\omega}\mathbb{R}$ *is bounded.*

Theorem 2 (W. Hurewicz [9]). *If X is a perfectly normal topological space then the following are equivalent.*

a. *X is a* $U_{\mathrm{fin}}(\mathcal{O}_{\mathrm{ctbl}}, \Gamma)$

b. *For every sequence* $\langle f_n : n \in \omega \rangle$ *of continuous real functions the family* $\{\langle f_n(x) : n \in \omega \rangle : x \in X\} \subseteq {}^{\omega}\mathbb{R}$ *is dominating.*

Note that the property $U_{\mathrm{fin}}(\mathcal{O}_{\mathrm{ctbl}}, \Gamma)$ of a topological space was introduced and investigated by K. Menger [10].

Proofs of both Theorems may be found, e.g., in L. Bukovský and J. Haleš [11].

A topological space X is a γ-**space** if every open ω-cover of X has a countable γ-subcover.

F. Gerlits and Z. Nagy [12] proved.

Theorem 3 (F. Gerlits and Z. Nagy [12]). *If X is a Tychonoff topological space then the following are equivalent:*

a. $\mathrm{C}_p(X)$ *is Fréchet.*

b. *X is a* γ*-space.*

c. *X is an* $\mathrm{S}_1(\Omega, \Gamma)$*-space.*

A topological space X has **countable strong fan tightness** if $A_n \subseteq X$ and $x \in \overline{A_n}$, $n \in \omega$ imply that there exists a sequence $\langle x_n : n \in \omega \rangle$ such that $x_n \in A_n$ and $x \in \overline{\{x_n : n \in \omega\}}$.

Theorem 4 (M. Sakai [13]). *A Tychonoff topological space has the covering property* $\mathrm{S}_1(\Omega, \Omega)$ *if and only if the topological space* $\mathrm{C}_p(X)$ *has countable strong fan tightness.*

That was M. Scheepers [2] who began the systematic study of the covering properties $\mathrm{G}(\Phi, \Psi)$ for $\mathrm{G} = \mathrm{S}_1, \mathrm{S}_{\mathrm{fin}}, U_{\mathrm{fin}}$, $\Phi, \Psi = \mathcal{O}, \Lambda, \Omega, \Gamma$.

The first use of upper semicontinuous functions in the study of covering properties was.

Theorem 5 (L. Bukovský [1]). *A topological space X is an* $\mathrm{S}_1(\Gamma, \Gamma)$*-space if and only if* $\mathrm{USC}_p(X)^+$ *satisfies the selection principle* $\mathrm{S}_1(\Gamma_{\mathbf{0}}, \Gamma_{\mathbf{0}})$.

Later on we succeeded to prove a general result.

Theorem 6 (L. Bukovský [14]). *Assume that* Φ *is one of the symbols* Ω *and* Γ, *and* Ψ *is one of the symbols* $\mathcal{O}$, Ω, Γ. *Then for any couple* $\langle \Phi, \Psi \rangle$ *different from* $\langle \Omega, \mathcal{O} \rangle$, *a topological space X is an* $\mathrm{S}_1(\Phi, \Psi)$*-space if and only if* $\mathrm{USC}_p(X)^+$ *satisfies the selection principle* $S_1(\Phi_{\mathbf{0}}, \Psi_{\mathbf{0}})$.

Similarly for $\mathrm{S}_{\mathrm{fin}}$ *and* U_{fin}.

To describe the selection principles of $C_p(X)$ we need different covers of the topological space X. If φ denotes one of the symbols o, ω or γ, then a φ-cover $\mathcal{U}$ is **shrinkable**, if there exists an open φ-cover $\mathcal{V}$ such that

$$(\forall V \in \mathcal{V})(\exists U_V \in \mathcal{U})\, \overline{V} \subseteq U_V. \tag{5}$$

The family $\{U_V : V \in \mathcal{V}\} \subseteq \mathcal{U}$ is a φ-cover as well. The family of all open shrinkable φ-covers of X will be denoted by $\Phi^{sh}(X)$.

Extending the result by L. Bukovský and J. Haleš [11] for $S_1(\Gamma^{sh}, \Gamma)$ we obtain.

Theorem 7 (L. Bukovský [14]). *Assume that* Φ *is one of the symbols* Ω, Γ *and* Ψ *is one of the symbols* $\mathcal{O}$, Ω, Γ. *Then for any couple* $\langle \Phi, \Psi \rangle$ *different from* $\langle \Omega, \mathcal{O} \rangle$ *a normal topological space* X *is an* $S_1(\Phi^{sh}, \Psi)$*-space if and only if* $C_p(X)$ *satisfies the selection principle* $S_1(\Phi_0, \Psi_0)$.

Similarly for S_{fin} *and* U_{fin}.

The next result is rather a folklore.

Theorem 8. *If* X *is a regular topological space, then for* $\Phi = \mathcal{O}, \Omega$ *and* $\Psi = \mathcal{O}, \Omega, \Gamma$ *we have* $\Phi^{sh} = \Phi$ *and therefore*

$$S_1(\Phi^{sh}, \Psi) \equiv S_1(\Phi, \Psi). \tag{6}$$

Corollary 9. *Let* X *be a normal topological space. Then for* $\Phi = \mathcal{O}, \Omega$ *and* $\Psi = \mathcal{O}, \Omega, \Gamma$ *the following are equivalent:*

a. $C_p(X)$ *satisfies the selection principle* $S_1(\Phi_0, \Psi_0)$

b. *The family* $USC(X)^+$ *satisfies the sequence selection principle* $S_1(\Phi_0, \Psi_0)$.

c. *The family* $C(X)$ *satisfies the sequence selection principle* $S_1(\Phi_0, \Psi_0)$.

Note that Theorem 4 is a special case of the Corollary.

4. The measurable covers and functions

Let $\mathcal{E} \subseteq \mathcal{P}(X)$ be as above, i.e. $\emptyset, X \in \mathcal{E}$. A real function $f \in {}^X\mathbb{R}$ is *upper* $\mathcal{E}$*-semimeasurable* if for every real $a \in \mathbb{R}$, the set $\{x \in X : f(x) < a\}$ belongs to $\mathcal{E}$. A real function $f \in {}^X\mathbb{R}$ is $\mathcal{E}$*-measurable* if for every reals $a < b \in \mathbb{R}$, including $a = -\infty, b = \infty$, the set $\{x \in X : a < f(x) < b\}$ belongs to $\mathcal{E}$. We denote by $USM(X, \mathcal{E})$ the set of all real upper $\mathcal{E}$-semimeasurable functions defined on X. Similarly, we denote by $M(X, \mathcal{E})$ the set of all real $\mathcal{E}$-measurable functions defined on X. Note that if $\mathcal{E}$ is a σ-algebra, then $M(X, \mathcal{E}) = USM(X, \mathcal{E})$.

If $\langle X, \tau \rangle$ is a topological space then $M(X, \tau) = C(X)$ and $USM(X, \tau) = USC(X)$.

Theorem 10 (L. Bukovský [15]). *Assume that* Φ *is one of the symbols* Ω *or* Γ, Ψ *is one of the symbols* $\mathcal{O}$, Ω *or* Γ, *and* $\langle \Phi, \Psi \rangle \neq \langle \Omega, \mathcal{O} \rangle$. *Let* $\mathcal{E}$ *be a family of subsets of a set* X, $\emptyset, X \in \mathcal{E}$. *If* $\Psi = \Gamma$, *we assume that* $\mathcal{E}$ *is closed under finite intersections.*

a. X *possesses the covering property* $S_1(\Phi(\mathcal{E}), \Psi(\mathcal{E}))$ *if and only if* $USM(X, \mathcal{E})^+$ *satisfies the selection principle* $S_1(\Phi_0, \Psi_0)$.

b. *Similarly for* S_{fin}.

c. *If* $\mathcal{E}$ *is closed under finite unions, then* X *possesses the covering property* $U_{fin}(\Phi(\mathcal{E}), \Psi(\mathcal{E}))$ *if and only if* $USM(X, \mathcal{E})^+$ *satisfies the selection principle* $U^*_{fin}(\Phi_0, \Psi_0)$.

If $\mathcal{E}$ *is a* σ*-algebra, then the family* $USM(X, \mathcal{E})^+$ *may be replaced by* $M(X, \mathcal{E})$.

For $\mathcal{E} = \tau$ you obtain Theorem 6. For $\mathcal{E} = BOREL$ you obtain some results of M. Scheepers and B. Tsaban [16].

5. Countable covers

A countable family may be considered as a sequence. If $\mathcal{U} = \langle U_n : n \in \omega \rangle$ is an (open) cover, then $\mathcal{V} = \langle \bigcup_{i \le n} U_i : n \in \omega \rangle$ is an (open) γ-cover. Some covering properties true for $\mathcal{U}$ remain true also for $\mathcal{V}$. E.g., sometimes choosing finite subsets of $\mathcal{V}$ is same as choosing finite subsets of $\mathcal{U}$.

If $\mathcal{U}$ is uncountable you cannot construct the γ-cover $\mathcal{V}$. Everything you can do is to construct an (open) ω-cover $\mathcal{V} = \{\bigcup \mathcal{W} : \mathcal{W} \subseteq \mathcal{U} \text{ finite}\}$. That is the essence of different behavior of countable and uncountable covers.

W. Just, A. Miller, M. Scheepers and P. Szeptycki [3] systematically studied the countable covering property $\mathrm{G}(\Phi_{\mathrm{ctbl}}, \Psi)$ for $\mathrm{G} = \mathrm{S}_1, \mathrm{S}_{\mathrm{fin}}, U_{\mathrm{fin}}, \Phi, \Psi = \mathcal{O}, \Lambda, \Omega, \Gamma$. They obtained several equivalences and as the result the Scheepers Diagram for Countable Covers (**Figure 1**).

Every countable covering property $\mathrm{G}(\Phi, \Psi)$, where Φ, Ψ are one of the symbols $\mathcal{O}, \Lambda, \Omega, \Gamma$ and G is one of the symbols $\mathrm{S}_1, \mathrm{S}_{\mathrm{fin}}, U_{\mathrm{fin}}$, is equivalent to some covering property in the Scheepers Diagram for Countable Covers. We do not know whether the thick arrow $\mathrm{S}_{\mathrm{fin}}(\Gamma, \Omega) \to U_{\mathrm{fin}}(\Gamma, \Omega)$ is reversible. All other arrows of the Diagram are at least consistently not reversible.

6. Sheepers' diagram for arbitrary covers

The Sheepers' Diagram for Countable Covers is valid only for countable covers. If we allow also uncountable covers, some equivalences, used for simplifying the diagram for countable covers, are generally false. We must make the corresponding corrections.

The first simplification consisted in equivalencies of $U_{\mathrm{fin}}(\Phi, \Psi)$ for different Φ. Not all of those equivalences are true for uncountable covers. Considering the countable covers, the following properties are equivalent for $\Psi = \mathcal{O}, \Lambda, \Omega, \Gamma$:

$$U_{\mathrm{fin}}(\mathcal{O}, \Psi) \equiv U_{\mathrm{fin}}(\Omega, \Psi) \equiv U_{\mathrm{fin}}(\Gamma, \Psi). \tag{7}$$

However, allowing uncountable covers we have only

$$U_{\mathrm{fin}}(\Omega, \Psi) \equiv U_{\mathrm{fin}}(\mathcal{O}, \Psi) \tag{8}$$

for $\Psi = \mathcal{O}, \Lambda, \Omega, \Gamma$ – see Example 11.

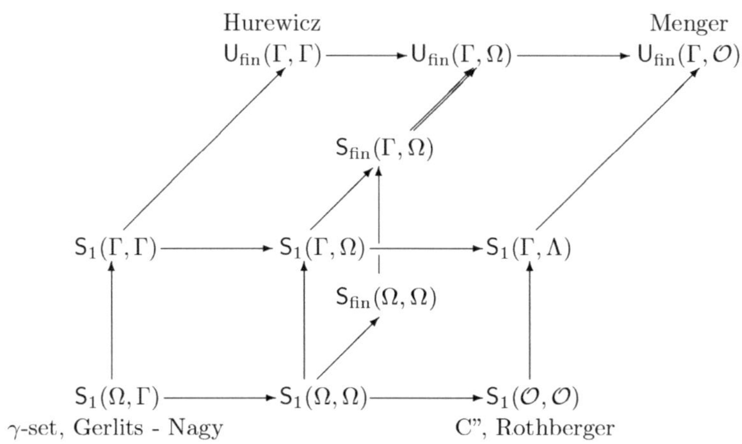

Figure 1.
Sheepers' diagram for countable covers.

In [3] the authors have shown that

$$\mathsf{S}_1(\Gamma, \Gamma) \equiv \mathsf{S}_{\text{fin}}(\Gamma, \Gamma). \tag{9}$$

The equivalence

$$\mathsf{S}_1(\Omega, \Gamma) \equiv \mathsf{S}_{\text{fin}}(\Omega, \Gamma) \tag{10}$$

remains true also for uncountable covers (an $\mathsf{S}_{\text{fin}}(\Omega, \Gamma)$-space is a γ-space). One can easily see that for $\Phi = \mathcal{O}, \Omega, \Gamma$ we have

$$\mathsf{S}_{\text{fin}}(\Omega, \mathcal{O}) \equiv \mathsf{S}_{\text{fin}}(\mathcal{O}, \mathcal{O}). \tag{11}$$

However, (11) is false at least consistently for $\mathsf{S}_{\text{fin}}(\Gamma, \Phi)$ and $U_{\text{fin}}(\Gamma, \Phi)$.

Example 11. *Assume that* $\Phi = \mathcal{O}, \Omega, \Gamma$. *Assume that* $\mathfrak{b} > \aleph_1$. *Then by Theorems 4.6, 4.7 of* [3] *and* (9), *the discrete space of cardinality* $\aleph_1$ *is* $\mathsf{S}_{\text{fin}}(\Gamma, \Phi)$ *and is not Lindelöf. On the other side,* $\mathsf{S}_{\text{fin}}(\mathcal{O}, \Phi)$ *and* $U_{\text{fin}}(\mathcal{O}, \Phi)$ *imply that X is Lindelöf.*

Example 12. *No topological space is* $\mathsf{S}_{\text{fin}}(\mathcal{O}, \Lambda)$. *So neither* $\mathsf{S}_{\text{fin}}(\mathcal{O}, \Omega)$ *nor* $\mathsf{S}_{\text{fin}}(\mathcal{O}, \Gamma)$. *Indeed, if X is an infinite Hausdorff topological space, fix a point* $a \in X$ *and an open neighborhood* $V \neq X$ *of a. For every* $x \notin V$ *take an open neighborhood* U_x *of x not*

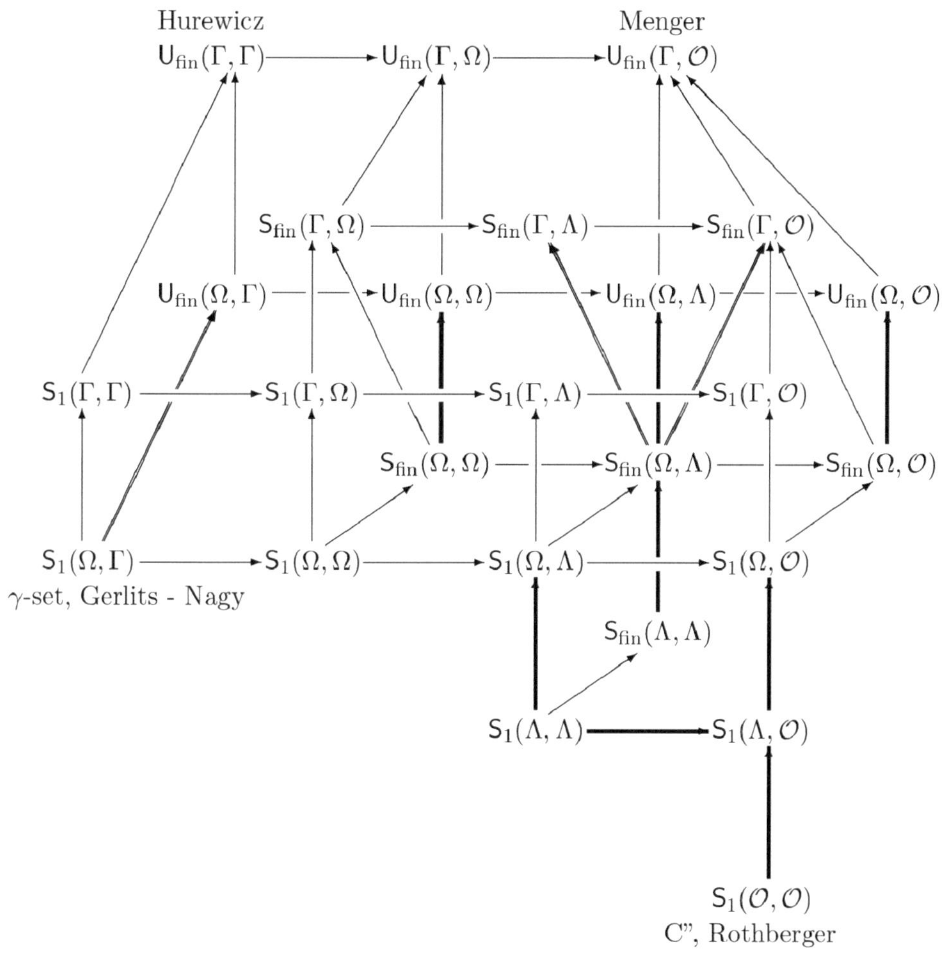

Figure 2.
Sheepers' Diagram for Arbitrary Infinite Covers.

containing a. Then $\mathcal{U} = \{U_x : x \in X \setminus V\} \cup \{V\}$ *is an open cover of X and no subcover of* $\mathcal{U}$ *is a* λ*-cover.*

Some covering properties are omitted, since they are equivalent with some others included in the **Figure 2**. We present those equivalences. Always the former member of an equivalence is included in the Diagram and the latter member is omitted.

First, take into accunt the equivalences (8)–(11). M. Scheepers [2] in Corollaries 5 and 6 has shown that $\mathsf{S}_{\mathrm{fin}}(\Gamma, \Lambda) \equiv U_{\mathrm{fin}}(\Gamma, \mathcal{O})$ and $\mathsf{S}_1(\Gamma, \Lambda) \equiv \mathsf{S}_1(\Gamma, \mathcal{O})$. Evidently $[U_{\mathrm{fin}}(\Omega, \Lambda) \equiv U_{\mathrm{fin}}(\Lambda, \Lambda) \equiv U_{\mathrm{fin}}(\mathcal{O}, \Lambda)$.

Taking in account all mentioned results, we obtain the diagram for arbitrary covers (**Figure 2**).

We do not know whether the eleven thick arrows of the Scheepers' Diagram (reversible for countable covers) are reversible for arbitrary covers. The other arrows are not reversible either by Figure 3 of [3] and corresponding results or by Example 11.

7. F_σ-subsets and covering properties

If $\mathcal{U}$ is a family of subsets of X, $A \subseteq X$, we set

$$\mathcal{U}|A = \{U \cap A : U \in \mathcal{U}\}. \tag{12}$$

Let $A \subseteq X$. If $U \subseteq A$ is open in the subspace topology, then there exists an open set $U^* \subseteq X$ such that $U = A \cap U^*$. Using the Axiom of Choice we choose one such set U^* for each set $U \subseteq A$ open in A. If $\mathcal{U}$ is an open (in the subspace topology) cover of A then we set $\mathcal{U}^* = \{U^* : U \in \mathcal{U}\}$. Then $\mathcal{U} = \mathcal{U}^*|A$. Note that $\mathcal{U}^*$ need not be a cover of X.

By Theorem 3.1 of [3] we have.

Theorem 13 (W. Just, A. Miller, P. Sczeptycki and M. Scheepers [3]). *Let* $\Phi, \Psi \in \{\mathcal{O}, \Lambda, \Omega, \Gamma\}$, G *being one of* $\mathsf{S}_1, \mathsf{S}_{\mathrm{fin}}, U_{\mathrm{fin}}$. *If a topological space X possesses the covering property* $\mathsf{G}(\Phi, \Psi)$, $F \subseteq X$ *is closed, then F with the subspace topology possesses the property* $\mathsf{G}(\Phi, \Psi)$ *as well. Moreover, if* $f : X \xrightarrow{onto} Y$ *is continuous, then Y possesses the property* $\mathsf{G}(\Phi, \Psi)$ *as well.*

Let $\mathcal{S}$ be a "topological" property of topological spaces, $\mathcal{E} \subseteq \mathcal{P}(X)$, $X \in \mathcal{E}$. We say that $\mathcal{S}$ is *hereditary for* $\mathcal{E}$ if assuming that X has the property $\mathcal{S}$, every $A \in \mathcal{E}$ endowed with the subspace topology has the property $\mathcal{S}$ as well. If $\mathcal{E} = \mathcal{P}(X)$ then we simply say that $\mathcal{S}$ is *hereditary on X*.

By Theorem 13, for any $\Phi, \Psi \in \{\mathcal{O}, \Lambda, \Omega, \Gamma\}$, G being one of $\mathsf{S}_1, \mathsf{S}_{\mathrm{fin}}, U_{\mathrm{fin}}$, the covering property $\mathsf{G}(\Phi, \Psi)$ is hereditary for closed subsets.

In [8], M. Scheepers proved that $\mathrm{add}(\mathsf{S}_1(\Gamma, \Gamma)) \geq \mathfrak{h}$. Since $\mathfrak{h} > \aleph_0$, by Theorem 13 we obtain.

Corollary 14. *The covering property* $\mathsf{S}_1(\Gamma, \Gamma)$ *is hereditary for* F_σ*-subsets of X.*

We prove the following result.

Theorem 15. *Let* $\Phi \in \{\mathcal{O}, \Omega, \Gamma\}$ *and* $\Psi \in \{\mathcal{O}, \Lambda, \Omega\}$. *Let* G *be one of* $\mathsf{S}_1, \mathsf{S}_{\mathrm{fin}}, U_{\mathrm{fin}}$. *The covering property* $\mathsf{G}(\Phi, \Psi)$ *is hereditary for* F_σ*-subsets of X.*

Proof: Let $F = \bigcup_n F_n$, where for each n the set F_n is closed and $F_n \subseteq F_{n+1}$. Let the countable sequence of open φ-covers of F be bijectively enumerated as

$$\langle \mathcal{U}_{n,m} : n, m \in \omega \rangle. \tag{13}$$

Then $\mathcal{U}_{n,m}|F_n$ is a φ-cover of F_n. Apply $\mathsf{S}_1(\Phi, \Psi)$ to the sequence $\langle \mathcal{U}_{n,m}|F_n : m \in \omega \rangle$ for every n. We obtain sequences $\langle V_{n,m} \in \mathcal{U}_{n,m}|F_n : m \in \omega \rangle$ such that every family $\{V_{n,m} : m \in \omega\}$ is a ψ-cover of F_n. Let $U_{n,m} \in \mathcal{U}_{n,m}$ be such that $V_{n,m} = U_{n,m} \cap F_n$.

One can easily see that the family $\{U_{n,m} : n, m \in \omega\}$ is a ψ-cover of F. Since $U_{n,m} \in \mathcal{U}_{n,m}$ for every n and m, we obtain that F possesses the covering property $\mathrm{S}_1(\Phi, \Psi)$.

For $\mathrm{G} = \mathrm{S}_{\mathrm{fin}}, U_{\mathrm{fin}}$ the proof is similar. □

8. σ-space

A topological space X is said to be a *σ-space*, if every G_δ-subset of X is an F_σ-set. Consequently, every Borel subset of a σ-space is an F_σ-set.

J. Haleš [4] proved.

Theorem 16. *Let X be a perfectly normal topological* $\mathrm{S}_1(\Gamma, \Gamma)$*-space. The covering property* $\mathrm{S}_1(\Gamma, \Gamma)$ *is hereditary on X if and only if X is a σ-space.*

We obtain an easy Corollary.

Corollary 17. *Let Φ be one of the symbols $\mathcal{O}, \Lambda, \Omega, \Gamma$. Let* G *be one of* $\mathrm{S}_1, \mathrm{S}_{\mathrm{fin}}$. *Assume that X is a perfectly normal topological space. If X possesses the covering property* $\mathrm{G}(\Phi, \Gamma)$ *and* $\mathrm{G}(\Phi, \Gamma)$ *is hereditary on X, then X is a σ-space.*

Proof: Note that

$$\mathrm{G}(\Phi, \Gamma) \to \mathrm{G}(\Gamma, \Gamma) \to \mathrm{S}_1(\Gamma, \Gamma), \tag{14}$$

(if $\mathrm{G} = \mathrm{S}_{\mathrm{fin}}$ use (9)) and $\Gamma \subseteq \Phi$. □

Following Haleš' proof of Theorem 16 we obtain.

Theorem 18. *Let X be a σ-space, Φ being one of the symbols $\mathcal{O}, \Lambda_{ctbl}, \Gamma$ and Ψ being one of $\mathcal{O}, \Lambda, \Omega, \Gamma$. Let* G *be one of* $\mathrm{S}_1, \mathrm{S}_{\mathrm{fin}}, U_{\mathrm{fin}}$. *Then* $\mathrm{G}(\Phi, \Psi)$ *is hereditary on X.*

Proof: Assume that X is a topological $\mathrm{G}(\Phi, \Psi)$-space, where $\Phi = \mathcal{O}, \Lambda_{ctbl}, \Gamma$ and $\Psi = \mathcal{O}, \Lambda, \Omega, \Gamma$. Assume also that X is a σ-space and $A \subseteq X$. Let $\langle \mathcal{U}_n : n \in \omega \rangle$ be a sequence of open (in the subspace topology) φ-covers of A.

If $\Phi = \mathcal{O}$ then $\mathcal{U}_n^* | B$ is an o-cover of $B = \bigcap_n \bigcup \mathcal{U}_n^* \supseteq A$. Since B is a Borel set, therefore an F_σ-set in X, by Theorem 15, the set B possesses the covering property $\mathrm{G}(\mathcal{O}, \Psi)$. So, for each $n \in \omega$ there exists a $U_n \in \mathcal{U}_n^* | B$ or finite set $\mathcal{W}_n \subseteq \mathcal{U}_n^* | B$ such that $\{U_n : n \in \omega\}$, or $\bigcup_n \mathcal{W}_n$, or $\{\bigcup \mathcal{W}_n : n \in \omega\}$ is a ψ-cover of B, respectively. Then $\{U_n : n \in \omega\} | A$, or $\bigcup_n \mathcal{W}_n | A$, or $\{\bigcup \mathcal{W}_n : n \in \omega\} | A$ is an open ψ-cover of A, respectively.

Let $\Phi = \Lambda_{ctbl}$. Since each $\mathcal{U}_n$ is countable, we can assume that $\{U_{n,m} : m \in \omega\}$ is a bijective enumeration of $\mathcal{U}_n$. The family $\mathcal{U}_{n,m} = \{U_{n,k} : k \geq m\}$ is a λ-cover of A. If we set $B = \bigcap_{n,m} \bigcup \mathcal{U}_{n,m}^* \supseteq A$ then each $\mathcal{U}_n^* | B$ is a λ-cover of B. Continue as above.

If $\Phi = \Gamma$ we can assume that each $\mathcal{U}_n$ is countable. For every $n \in \omega$ let $\{U_{n,m} : m \in \omega\}$ be a bijective enumeration of $\mathcal{U}_n$. Then $\mathcal{U}_n^* | B$ is a γ-cover of $B = \bigcap_n \bigcup_k \cap_{m \geq k} U_{n,m}^* \supseteq A$. The set B is Borel, therefore F_σ. By Theorem 15, B is an $\mathrm{G}(\Gamma, \Psi)$-space. Continue as above. □

For $\Phi = \Psi = \Gamma$, $\mathrm{G} = \mathrm{S}_1$, we obtain one implication of the Haleš' Theorem 16.

Corollary 19. *Let Φ be one of the symbols $\mathcal{O}, \Lambda_{ctbl}, \Gamma$. Let* G *be one of* $\mathrm{S}_1, \mathrm{S}_{\mathrm{fin}}$. *The covering property* $\mathrm{G}(\Phi, \Gamma)$ *is hereditary on a perfectly normal topological space X if and only if X is a σ-space.*

Corollary 20. *Let Φ be one of the symbols $\mathcal{O}, \Lambda_{ctbl}, \Gamma$. Let* G *be one of* $\mathrm{S}_1, \mathrm{S}_{\mathrm{fin}}$. *Assume that a perfectly normal topological space X possesses the covering property* $\mathrm{G}(\Phi, \Gamma)$. *Then X is hereditary* $\mathrm{G}(\Phi, \Gamma)$ *if and only if X is hereditary* $\mathrm{S}_1(\Gamma, \Gamma)$.

9. Remarks

I have obtained a short time for writing this paper. So, I have no time to collect all results known before proving Theorems 5, 6, 7, 10, 15 and 18. For a partial presentation of such known results see [14, 15].

Classification

2010 MSC: 54C35, 54C20, 54D55

Author details

Lev Bukovský[†]
Institute of Mathematics, P.J. Šafárik University, Košice, Slovakia

*Address all correspondence to: lev.bukovsky@upjs.sk

† Deceased.

References

[1] Bukovský L., *On* wQN$_*$ *and* wQN* *spaces,* Topology Appl. **156** (2008), 24–27.

[2] Scheepers M., *Combinatorics of open covers* I*: Ramsey theory,* Topology Appl. **69** (1996), 31–62.

[3] Just W., Miller A.W., Scheepers M. and Szeptycki P.J., *Combinatorics of open covers II,* Topology Appl. **73** (1996), 241–266.

[4] Haleš J., *On Scheepers' Conjecture,* Acta Univ. Carolinae – Math. Phys. **46** (2005), 27–31.

[5] Engelking R., General Topology, Monografie Matematyczne **60**, Warszawa 1977, revised edition Heldermann Verlag, Berlin 1989.

[6] Blass A. Combinatorial cardinal characteristics of the continuum. In: Foreman M, Kanamori A, editors. Handbook of Set Theory. Dordrecht: Springer; 2010. pp. 24-27

[7] van Douwen E. The integers and topology. In: Kunen K, Vaughan J, editors. Handbook of Set-Theoretic Topology. North-Holland; 1984. pp. 111-167

[8] Scheepers M., *A sequential property of* $C_p(X)$ *and a covering property of Hurewicz,* Proc. Amer. Math. Soc. **125** (1997), 2789–2795.

[9] Hurewicz W., *Über Folgen stetiger Funktionen,* Fund. Math. **9** (1927), 193–204.

[10] Menger K., *Einige Überdeckungssatze der Punktmengenlehre,* Sitzungsberichte. Abt. 2a, Mathematik, Astronomie, Physik, Meteorologie und Mechanik (Wiener Akademie) **133** (1924), 421–444.

[11] Bukovský L. and Haleš J., *On Hurewicz Properties,* Topology Appl. **132** (2003), 71-79.

[12] Gerlits F. and Nagy Z., *Some properties of* C*(X), part I,* Topology Appl. **14** (1982), 151–161.

[13] Sakai M., *Property* C" *and function spaces,* Proc. Amer. Math. Soc. **104** (1988), 917–919.

[14] Bukovský L., *Selection principle* S_1 *and combinatorics of open covers,* Topology Appl. **258** (2019), 239–252.

[15] Bukovský L., *Measurable Functions and Covering Properties,* Topology Appl. **258** (2021), 107787.

[16] Scheepers M. and Tsaban B., *The combinatorics of Borel covers,* Topology Appl. **121** (2002), 357–382.

Section 4

Combinatoral Topology and Descompoibilities to Shellability

Chapter 6

Vertex Decomposability of Path Complexes and Stanley's Conjectures

Seyed Mohammad Ajdani and Francisco Bulnes

Abstract

Monomials are the link between commutative algebra and combinatorics. With a simplicial complex Δ, one can associate two square-free monomial ideals: the Stanley-Reisner ideal I_Δ whose generators correspond to the non-face of Δ, or the facet ideal $I(\Delta)$ that is a generalization of edge ideals of graphs and whose generators correspond to the facets of Δ. The facet ideal of a simplicial complex was first introduced by Faridi in 2002. Let G be a simple graph. The edge ideal $I(G)$ of a graph G was first considered by R. Villarreal in 1990. He studied algebraic properties of $I(G)$ using a combinatorial language of G. In combinatorial commutative algebra, one can attach a monomial ideal to a combinatorial object. Then, algebraic properties of this ideal are studied using combinatorial properties of combinatorial object. One of interesting problems in combinatorial commutative algebra is the Stanley's conjectures. The Stanley's conjectures are studied by many researchers. Let R be a $\mathbb{N}^n$-graded ring and M a $\mathbb{Z}^n$-graded R-module. Then, Stanley conjectured that $\text{depth}(M) \leq \text{sdepth}(M)$. He also conjectured that each Cohen-Macaulay simplicial complex is partition-able. In this chapter, we study the relation between vertex decomposability of some simplicial complexes and Stanley's conjectures.

Keywords: vertex decomposable, simplicial complex, Matroid, path

1. Introduction

Let $R = K[x_1, \dots, x_n]$, where K is a field. Fix an integer $n \geq t \geq 2$ and let G be a directed graph. A sequence $x_{i_1}, \dots, x_{i_t}$ of distinct vertices is called a path of length t if there are $t-1$ distinct directed edges $e_1, \dots, e_{t-1}$ where e_j is a directed edge from x_{i_j} to $x_{i_{j+1}}$. Then, the path ideal of G of length t is the monomial ideal $I_t(G) = (x_{i_1} \dots x_{i_t} : x_{i_1}, \dots, x_{i_t}$ is a path of length t in G) in the polynomial ring $R = K[x_1, \dots, x_n]$. The distance $d(x, y)$ of two vertices x and y of a graph G is the length of the shortest path from x to y. The path complex $\Delta_t(G)$ is defined by

$$\Delta_t(G) = \langle \{x_{i_1}, \dots, x_{i_t}\} : x_{i_1}, \dots, x_{i_t} \quad \text{is a path of length } t \text{ in } G \rangle.$$

IntechOpen

Path ideals of graphs were first introduced by Conca and De Negri [1, 2] in the context of monomial ideals of linear type. Recently, the path ideal of cycles has been extensively studied by several mathematicians. In [3], it is shown that $I_2(C_n)$ is sequentially Cohen-Macaulay, if and only if, $n = 3$ or $n = 5$. Generalizing this result, in [4], it is proved that $I_t(C_n)$, $(t > 2)$, is sequentially Cohen-Macaulay, if and only if $n = t$ or $n = t + 1$ or $n = 2t + 1$. Also, the Betti numbers of the ideal $I_t(C_n)$ and $I_t(L_n)$ is computed explicitly in [5]. In particular, it has been shown that [6, 7]:

Theorem 1.1 ([1, Corollary 5.15]). *Let n, t, p and d be integers such that $n \geq t \geq 2$, $n = (t+1)p + d$, where $p \geq 0$ and $0 \leq d < (t+1)$. Then,*

i. *The projective dimension of the path ideal of a graph cycle C_n or line L_n is given by,*

$$\mathrm{pd}(I_t(C_n)) = \begin{cases} 2p, & d \neq 0 \\ 2p-1, & d = 0 \end{cases} \qquad \mathrm{pd}(I_t(L_n)) = \begin{cases} 2p-1, & d \neq t \\ 2p, & d = t \end{cases} \tag{1}$$

ii. *The regularity of the path ideal of a graph cycle C_n or line L_n is given by,*

$$\begin{aligned} \mathrm{reg}(I_t(C_n)) &= (t-1)p + d + 1 \\ \mathrm{reg}(I_t(L_n)) &= \begin{cases} p(t-1)+1, & d < t \\ p(t-1)+t, & d = t \end{cases} \end{aligned} \tag{2}$$

In [8] it has been shown that $\Delta_t(G)$ is a simplicial tree if G is a rooted tree and $t \geq 2$. One of interesting problems in combinatorial commutative algebra is the Stanley's conjectures. The Stanley's conjectures are studied by many researchers. Let R be a $\mathbb{N}^n$-graded ring and M a $\mathbb{Z}^n$-graded R-module. Then, Stanley [9] conjectured that

$$\mathrm{depth}(M) \leq \mathrm{sdepth}(M) \tag{3}$$

He also conjectured in [10] that each Cohen-Macaulay simplicial complex is partitionable. Herzog, Soleyman Jahan, and Yassemi in [11–14] showed that the conjecture about partitionability is a special case of the Stanley's first conjecture. In this chapter, we first study algebraic properties of $\Delta_t(C_n)$. In Section 1, we recall some definitions and results, which will be needed later. In Section 2, for all $t > 2$ we show that the following conditions are equivalent:

i. $\Delta_t(C_n)$ is matroid;

ii. $\Delta_t(C_n)$ is vertex decomposable;

iii. $\Delta_t(C_n)$ is shellable;

iv. $\Delta_t(C_n)$ is Cohen-Macaulay;

v. $n = t$ or $t + 1$.

(see Theorem 2.6).

In Section 3, for all $t \geq 2$ we show that $\Delta_t(G)$ is vertex decomposable if and only if $G = H(p, n, q)$ or $G = H(p, n)$. In Section 4, vertex decomposability path complexes of Dynkin graphs are shown. In Section 5 as an application of our results, we show that if $n = t$ or $t + 1$ then $\Delta_t(C_n)$ is partitionable and Stanley's conjecture holds for $K[\Delta_t(C_n)]$ and $K[\Delta_t(G)]$, where $G = H(p, n, q)$ or $G = H(p, n)$.

2. Preliminaries

In this section, we recall some definitions and results which will be needed later.

Definition 2.1. *A simplicial complex* Δ over a set of vertices $V = \{x_1, \dots, x_n\}$, is a collection of subsets of V, with the property that:

a. $\{x_i\} \in \Delta$, for all i;

b. if $F \in \Delta$, then all subsets of F are also in Δ (including the empty set).

An element of Δ is called a *face* of Δ and complement of a face F is $V \backslash F$ and it is denoted by F^c. Also, the complement of the simplicial complex $\Delta = \langle F_1, \dots, F_r \rangle$ is $\Delta^c = \langle F_1^c, \dots, F_r^c \rangle$. The *dimension* of a face F of Δ, dim F, is $|F| - 1$ where $|F|$ is the number of elements of F and dim $\emptyset = -1$. The faces of dimensions 0 and 1 are called *vertices* and *edges*, respectively. A *non-face* of Δ is a subset F of V with $F \notin \Delta$. We denote by $\mathcal{N}(\Delta)$, the set of all minimal non-faces of Δ. The maximal faces of Δ under inclusion are called *facets* of Δ. The *dimension* of the simplicial complex Δ, dim Δ, is the maximum of dimensions of its facets. If all facets of Δ have the same dimension, then Δ is called *pure*.

Let $\mathcal{F}(\Delta) = \{F_1, \dots, F_q\}$ be the facet set of Δ. It is clear that $\mathcal{F}(\Delta)$ determines Δ completely and we write $\Delta = \langle F_1, \dots, F_q \rangle$. A simplicial complex with only one facet is called a *simplex*. A simplicial complex Γ is called a *subcomplex* of Δ, if $\mathcal{F}(\Gamma) \subset \mathcal{F}(\Delta)$.

For $v \in V$, the subcomplex of Δ obtained by removing all faces $F \in \Delta$ with $v \in F$ is denoted by $\Delta \backslash v$. That is,

$$\Delta \backslash v = \langle F \in \Delta : \quad v \notin F \rangle. \tag{4}$$

The *link* of a face $F \in \Delta$, denoted by $\text{link}_\Delta(F)$, is a simplicial complex on V with the faces, $G \in \Delta$ such that, $G \cap F = \emptyset$ and $G \cup F \in \Delta$. The link of a vertex $v \in V$ is simply denoted by $\text{link}_\Delta(v)$.

$$\text{link}_\Delta(v) = \{F \in \Delta : \quad v \notin F, \quad F \cup \{v\} \in \Delta\}. \tag{5}$$

Let Δ be a simplicial complex over n vertices $\{x_1, \dots, x_n\}$. For $F \subset \{x_1, \dots, x_n\}$, we set:

$$x_F = \prod_{x_i \in F} x_i. \tag{6}$$

We define the *facet ideal* of Δ, denoted by $I(\Delta)$, to be the ideal of S generated by $\{x_F : \ F \in \mathcal{F}(\Delta)\}$. The *non-face ideal* or the *Stanley-Reisner ideal* of Δ, denoted by I_Δ, is the ideal of S generated by square-free monomials $\{x_F : \ F \in \mathcal{N}(\Delta)\}$. Also, we call $K[\Delta] := S/I_\Delta$ the *Stanley-Reisner ring* of Δ.

Definition 2.2. A simplicial complex Δ on $\{x_1, \dots, x_n\}$ is said to be a matroid if, for any two facets F and G of Δ and any $x_i \in F$, there exists a $x_j \in G$ such that $(F \backslash \{x_i\}) \cup \{x_j\}$ is a facet of Δ.

Definition 2.3. A simplicial complex Δ is recursively defined to be *vertex decomposable*, if it is either a simplex, or else has some vertex v so that,

a. Both $\Delta\backslash v$ and $\text{link}_\Delta(v)$ are vertex decomposable, and

b. No face of $\text{link}_\Delta(v)$ is a facet of $\Delta\backslash v$.

A vertex v which satisfies in condition (b) is called a *shedding vertex*.

Definition 2.4. A simplicial complex Δ is *shellable*, if the facets of Δ can be ordered $F_1, \dots, F_s$ such that, for all $1 \le i < j \le s$, there exists some $v \in F_j\backslash F_i$ and some $l \in \{1, \dots, j-1\}$ with $F_j\backslash F_l = \{v\}$.

A simplicial complex Δ is called disconnected, if the vertex set V of Δ is a disjoint union $V = V_1 \cup V_2$ such that no face of Δ has vertices in both V_1 and V_2. Otherwise, Δ is connected. It is well known that

$$\text{Matroid} \Rightarrow \text{vertex decomposable} \Rightarrow \text{shellable} \Rightarrow \text{Cohen-Macaulay}$$

Definition 2.5. Given a simplicial complex Δ on V, we define $\Delta^\vee$, the *Alexander dual* of Δ, by

$$\Delta^\vee = \{V\backslash F : \quad F \notin \Delta\}. \tag{7}$$

It is known that for the complex Δ one has $I_{\Delta^\vee} = I(\Delta^c)$. Let $I \neq 0$ be a homogeneous ideal of S and $\mathbb{N}$ be the set of non-negative integers. For every $i \in \mathbb{N} \cup \{0\}$, one defines:

$$t_i^S(I) = \max\left\{ j : \quad \beta_{i,j}^S(I) \neq 0 \right\} \tag{8}$$

where $\beta_{i,j}^S(I)$ is the i,j-th graded Betti number of I as an S-module. The *Castelnuovo-Mumford regularity* of I is given by

$$\text{reg}(I) = \sup\{t_i^S(I) - i : \quad i \in Z\}. \tag{9}$$

We say that the ideal I has a *d-linear resolution*, if I is generated by homogeneous polynomials of degree d and $\beta_{i,j}^S(I) = 0$, for all $j \neq i + d$ and $i \ge 0$. For an ideal that has a d-linear resolution, the Castelnuovo-Mumford regularity would be d. If I is a graded ideal of S, then we write (I_d) for the ideal generated by all homogeneous polynomials of degree d belonging to I.

Definition 2.6. A graded ideal I is componentwise linear if (I_d) has a linear resolution for all d.

Also, we write $I_{[d]}$ for the ideal generated by the squarefree monomials of degree d belonging to I.

Definition 2.7. A graded S-module M is called *sequentially Cohen-Macaulay* (over K), if there exists a finite filtration of graded S-modules,

$$0 = M_0 \subset M_1 \subset \cdots \subset M_r = M \tag{10}$$

such that each M_i/M_{i-1} is Cohen-Macaulay, and the Krull dimensions of the quotients are increasing:

$$\dim(M_1/M_0) < \dim(M_2/M_1) < \cdots < \dim(M_r/M_{r-1}). \tag{11}$$

The Alexander dual allows us to make a bridge between (sequentially) Cohen-Macaulay ideals and (componetwise) linear ideals.

Definition 2.8 (Alexander duality). For a square-free monomial ideal $I = (M_1, \dots, M_q) \subset S = K[x_1, \dots, x_n]$, the *Alexander dual* of I, denoted by $I^\vee$, is defined to be:

$$I^\vee = P_{M_1} \cap \cdots \cap P_{M_q} \tag{12}$$

where, P_{M_i} is prime ideal generated by $\{x_j : \quad x_j | M_i\}$.

Theorem 2.1 ([7, Proposition 8.2.20], [5, Theorem 3]). *Let I be a square-free monomial ideal in $S = K[x_1, \dots, x_n]$.*

i. *The ideal I is componentwise linear ideal if and only if $S/I^\vee$ is sequentially Cohen-Macaulay.*

ii. *The ideal I has a q-linear resolution if and only if $S/I^\vee$ is Cohen-Macaulay of dimension $n - q$.*

Remark 2.1. Two special cases, we will be considering in this paper, are when G is a cycle C_n, or a line graph L_n on vertices $\{x_1, \dots, x_n\}$ with edges

$$E(C_n) = \{\{x_1, x_2\}, \{x_2, x_3\}, \dots, \{x_{n-1}, x_n\}, \{x_n, x_1\}\};$$
$$E(L_n) = \{\{x_1, x_2\}, \{x_2, x_3\}, \dots, \{x_{n-1}, x_n\}\}$$

Remark 2.2. All Cohen-Macaulay simplicial complexes of positive dimension are connected.

3. Vertex decomposability path complexes of cycles

As the main result of this section, it is shown that $\Delta_t(C_n)$ is matroid, vertex decomposable, shellable, and Cohen-Macaualay if and only if $n = t$ or $n = t + 1$. For the proof, we shall need the following lemmas and propositions.

Lemma 3.1. *Let $\Delta_t(Pn)$ be a simplicial complex on the path $Pn = \{x_1, \dots, x_n\}$ and $2 \le t \le n$. Then $\Delta_t(Pn)$ is vertex decomposable.*

Proof. If $t = n$, then $\Delta_n(Pn)$ is a simplex which is vertex decomposable. Let $2 \le t < n$ then one has

$$\Delta_t(Pn) = \langle \{x_1, \dots, x_t\}, \{x_2, \dots, x_{t+1}\}, \dots, \{x_{n-t+1}, \dots, x_n\} \rangle. \tag{13}$$

So $\Delta_t(Pn) \backslash x_n = \langle \{x_1, \dots, x_t\}, \{x_2, \dots, x_{t+1}\}, \dots, \{x_{n-t}, \dots, x_{n-1}\} \rangle$. Now, we use induction on the number of vertices of Pn and by induction hypothesis $\Delta_t(Pn) \backslash x_n$ is vertex decomposable. On the other hand, it is clear that $link_{\Delta_t(Pn)}\{x_n\} = \langle \{x_{n-t+1}, \dots, x_{n-1}\} \rangle$. Thus, $link_{\Delta_t(Pn)}\{x_n\}$ is a simplex which is not a facet of $\Delta_t(Pn) \backslash x_n$. Therefore, $\Delta_t(Pn)$ is vertex decomposable.

Lemma 3.2. *Let $\Delta_2(C_n)$ be a simplicial complex on $\{x_1, \dots, x_n\}$. Then $\Delta_2(C_n)$ is vertex decomposable.*

Proof. Since $\Delta_2(C_n) = \langle \{x_1, x_2\}, \{x_2, x_3\}, \dots, \{x_{n-1}, x_n\}, \{x_n, x_1\} \rangle$ then we have

$$\Delta_2(C_n) \backslash x_n = \langle \{x_1, , x_2\}, \{x_2, x_3\}, \dots, \{x_{n-2}, x_{n-1}\} \rangle. \tag{14}$$

By Lemma 2.1 $\Delta_2(C_n) \backslash x_n$ is vertex decomposable. Also, it is trivial that $\text{link}_{\Delta_2(C_n)}\{x_n\} = \langle \{x_{n-1}\}, \{x_1\} \rangle$ is vertex decomposable and no face of $\text{link}_{\Delta_2(C_n)}\{x_n\}$ is a facet of $\Delta_2(C_n)\backslash x_n$. Therefore, $\Delta_2(C_n)$ is vertex decomposable. □

Lemma 3.3. *Let $\Delta_t(C_n)$ be a simplicial complex on $\{x_1, \dots, x_n\}$ and $3 \le t \le n-2$. Then $\Delta_t(C_n)$ is not Cohen-Macaulay.*

Proof. It suffices to show that $I_{\Delta_t(C_n)^\vee}$ has not a linear resolution. Since $I_{\Delta_t(C_n)^\vee} = I(\Delta_t(C_n)^c)$, then one can easily check that $I_{\Delta_t(C_n)^\vee} = I_{n-t}(C_n)$. By Theorem 1.1 we have

$$\text{reg}\left(I_{\Delta_t(C_n)^\vee}\right) = (n-t-1)p + d + 1. \tag{15}$$

Since $3 \le t \le n-2$ then one has $reg\left(I_{\Delta_t(C_n)^\vee}\right) \neq n-t$ and by Theorem 2.1 $\Delta_t(C_n)$ is not Cohen-Macaulay.

Proposition 3.1. *Let $\Delta_t(C_n)$ be a simplicial complex on $\{x_1, \dots, x_n\}$ and $t \ge 3$. Then $\Delta_t(C_n)$ is vertex decomposable if and only if $n = t$ or $t+1$.*

Proof. By Lemma 3.3, it suffices to show that if $n = t$ or $t+1$, then $\Delta_t(C_n)$ is vertex decomposable. If $n = t$, then $\Delta_n(C_n)$ is a simplex which is vertex decomposable.

If $t = n-1$, then we have

$$\Delta_{n-1}(C_n) = \langle \{x_1, \dots, x_{n-1}\}, \{x_2, \dots, x_n\}, \{x_3, \dots, x_n, x_1\}, \dots, \{x_n, x_1, \dots, x_{n-2}\} \rangle.$$

Now, we use induction on the number of vertices of C_n and show that $\Delta_{n-1}(C_n)$ is vertex decomposable. It is clear that $\Delta_{n-1}(C_n)\backslash x_n = \langle \{x_1, \dots, x_{n-1}\} \rangle$ is a simplex, which is vertex decomposable.

On the other hand,

$$\text{link}_{\Delta_{n-1}(C_n)}\{x_n\} = \langle \{x_1, \dots, x_{n-2}\}, \dots, \{x_{n-1}, x_1, \dots, x_{n-3}\} \rangle = \Delta_{n-2}(C_{n-1}). \tag{16}$$

By induction hypothesis $\text{link}_{\Delta_{n-1}(C_n)}\{x_n\}$ is vertex decomposable. It is easy to see that no face of $\text{link}_{\Delta_{n-1}(C_n)}\{x_n\}$ is a facet of $\Delta_{n-1}(C_n)\backslash x_n$. Therefore, $\Delta_{n-1}(C_n)$ is vertex decomposable.

Proposition 3.2. *$\Delta_2(C_n)$ is a matroid if and only if $n = 3$ or 4.*

Proof. If $n = 3$ or 4, then it is easy to see that $\Delta_2(C_n)$ is a matroid. Now, we prove the converse. It suffices to show that $\Delta_2(C_n)$ is not a matroid for all $n \ge 5$. We consider two facets $\{x_1, x_2\}$ and $\{x_{n-1}, x_n\}$. Then, we have $(\{x_1, x_2\}\backslash\{x_1\}) \cup \{x_{n-1}\} = \{x_2, x_{n-1}\}$ and $(\{x_1, x_2\}\backslash\{x_1\}) \cup \{x_n\} = \{x_2, x_n\}$. Since $\{x_2, x_{n-1}\}$ and $\{x_2, x_n\}$ are not the facets of $\Delta_2(C_n)$. So $\Delta_2(C_n)$ is not matroid for all $n \ge 5$.

For the simplicial complexes, one has the following implication:

$$\textbf{Matroid} \Rightarrow \textbf{vertex decomposable} \Rightarrow \textbf{shellable} \Rightarrow \textbf{Cohen} - \textbf{Macaulay}$$

Note that these implications are strict, but by the following theorem, for path complexes, the reverse implications are also valid.

Theorem 3.1. *Let $t \ge 3$. Then the following conditions are equivalent:*

i. *$\Delta_t(C_n)$ is matroid;*

ii. *$\Delta_t(C_n)$ is vertex decomposable;*

iii. *$\Delta_t(C_n)$ is shellable;*

iv. *$\Delta_t(C_n)$ is Cohen-Macaulay;*

v. *$n = t$ or $t+1$.*

Proof. (i)⇒ (ii), (ii)⇒ (iii) and (iii)⇒ (iv) is well known.
(iv)⇒ (v): Follows from Lemma 3.3 and Proposition 2.1.
(v)⇒ (i): If $n = t$, then $\Delta_t(C_n)$ is a simplex which is a matroid.
If $n = t + 1$, then

$$\Delta_t(C_n) = \langle \{x_1, \dots, x_t\}, \{x_2, \dots, x_{t+1}\}, \{x_3, \dots, x_{t+1}, x_1\}, \dots, \{x_{t+1}, x_1, \dots, x_{t-1}\} \rangle.$$

For any two facets F and G of $\Delta_t(C_n)$ one has $|F \cap G| = t - 1$. We claim that for any two facets F and G of $\Delta_t(C_n)$ and any $x_i \in F$, there exists a $x_j \in G$ such that $(F \backslash \{x_i\}) \cup \{x_j\}$ is a facet of $\Delta_t(C_n)$. We have to consider two cases. If $x_i \in F$ and $x_i \notin G$, then we choose $x_j \in G$ such that $x_j \notin F$. Thus, $(F \backslash \{x_i\}) \cup \{x_j\} = G$ which is a facet of $\Delta_t(C_n)$.

For other case, if $x_i \in F$ and $x_i \in G$, then we choose $x_j \in G$ such that x_j is the same x_i. Therefore, $(F \backslash \{x_i\}) \cup \{x_i\} = F$ is a facet of $\Delta_t(C_n)$, which completes the proof.

4. Vertex decomposability path complexes of trees

As the main result of this section, for all $t \geq 2$, we characterize all such trees whose $\Delta_t(G)$ is vertex decomposable. Let $H(p, n, q)$ denote the double starlike tree obtained by attaching p pendant vertices to one pendant vertex of the path Pn and q pendant vertices to the other pendant vertex of Pn. Also, let $H(p, n)$ be graph obtained by attaching p pendant vertices to one pendant vertex of the path Pn.

Remark 4.1. Let $Pn = \{x_1, \dots, x_n\}$ be a path on vertices $\{x_1, \dots, x_n\}$ and $H(2, n)$ be a graph obtained by attaching two pendant vertices to pendant vertex x_n. Then, $\Delta_t(H(2, n))$ is vertex decomposable for all $t \geq 2$.

Proof. By Lemma 3.1 proof is trivial.

Proposition 4.1. *Let $Pn = \{x_1, \dots, x_n\}$ be a path on vertices $\{x_1, \dots, x_n\}$ and $H(p, n)$ be a graph obtained by attaching p pendant vertices to pendant vertex x_n. Then $\Delta_t(H(p, n))$ is vertex decomposable for all $t \geq 2$.*

Proof. We prove the proposition by induction on p the number of pendant vertices to pendant vertex x_n of Pn. If $p = 0$ or 1, then $H(p, n)$ is a path and by Lemma 2.1 $\Delta_t(H(p, n))$ is vertex decomposable. If $p = 2$, then by remark 4.1 $\Delta_t(H(p, n))$ is vertex decomposable. Now, let $p > 2$ and $\{y_1, \dots, y_p\}$ be p pendant vertices to pendant vertex x_n of Pn, then one has

$$H(p, n) \backslash \{y_1\} = H(p - 1, n) \tag{17}$$

and

$$\Delta_t(H(p, n)) \backslash \{y_1\} = \Delta_t(H(p - 1, n)). \tag{18}$$

Therefore by induction hypothesis $\Delta_t(H(p - 1, n))$ is vertex decomposable. So $\Delta_t(H(p, n)) \backslash \{y_1\}$ is vertex decomposable. If $t = 3$, then we have

$$\mathrm{link}_{\Delta_3(H(p,n))}\{y_1\} = \left\langle \{x_{n-1}, x_n\}, \{y_2, x_n\}, \dots, \{y_p, x_n\} \right\rangle \tag{19}$$

It is easy to see that $\mathrm{link}_{\Delta_3(H(p,n))}\{y_1\}$ is vertex decomposable and y_1 is a shedding vertex. If $t = 2$ or $t > 3$, one has

$$\mathrm{link}_{\Delta_t(H(p,n))}\{y_1\} = \langle \{x_{n-t+2}, \dots, x_n\} \rangle. \tag{20}$$

thus $\text{link}_{\Delta_t(H(p,n))}\{y_1\}$ is a simplex, which is not a facet of $\Delta_t(H(p,n))\backslash\{y_1\}$, therefore $\Delta_t(H(p,n))$ is vertex decomposable.

Lemma 4.1. *Let $p = 2$ and $q \geq 2$, Then $\Delta_t(H(2,n,q))$ is vertex decomposable for all $2 \leq t \leq n+2$.*

Proof. Let $H(2,n,q)$ denote the double starlike tree obtained by attaching two pendant vertices $\{y_1,y_2\}$ to pendant vertex x_1 of path Pn and $\{y'_1, \ldots, y'_q\}$ be pendant vertices to pendant vertex x_n of Pn. So by proposition 3.2 $\Delta_t(H(2,n,q))\backslash\{y_1\}$ is vertex decomposable. Now, we prove that $\text{link}_{\Delta_t(H(2,n,q))}\{y_1\}$ is vertex decomposable. If $t = 3$, then $\text{link}_{\Delta_3(H(2,n,q))}\{y_1\} = \langle\{x_1,x_2\},\{x_1,y_2\}\rangle$ which is vertex decomposable. If $t = n+2$, then

$$\text{link}_{\Delta_{n+2}(H(2,n,q))}\{y_1\} = \left\langle\{x_1, \ldots, x_n, y'_1\}, \{x_1, \ldots, x_n, y'_2\}, \ldots, \{x_1, \ldots, x_n, y'_q\}\right\rangle.$$

It is easy to see that $\text{link}_{\Delta_{n+2}(H(2,n,q))}\{y_1\}$ is vertex decomposable. If $t = 2$ or $4 \leq t \leq n+1$, then we have $\text{link}_{\Delta_t(H(2,n,q))}\{y_1\} = \langle\{x_1, \ldots, x_{t-1}\}\rangle$. Thus, $\text{link}_{\Delta_t(H(2,n,q))}\{y_1\}$ is a simplex, which is vertex decomposable. It is clear that y_1 is a shedding vertex. □

Proposition 4.2. *Let Q_1, Q_2 be two paths of maximum length k in tree G and y be a leaf of G such that $y \in Q_1 \cap Q_2$, $|Q_1 \cap Q_2| = L$. Then $\Delta_t(G)$ is not vertex decomposable.*

Proof. Suppose $Q_1 = \{y_1, y_2, \ldots, y_{k-L}, x_1, x_2, \ldots, x_{L-1}, y\}$ and. $Q_2 = \{y'_1, y'_2, \ldots, y'_{k-L}, x_1, x_2, \ldots, x_{L-1}, y\}$ be two paths of length k in G such that $Q_1 \cap Q_2 = \{x_1, x_2, \ldots, x_{L-1}, y\}$ and $deg(y) = 1$. Since $\text{link}_{\Delta_k(G)}\{x_1, \ldots, x_{L-1}, y\}$ is disconnected and pure of positive dimension. By remark 1.11 $\Delta_k(G)$ is not Cohen-Macaulay and hence, $\Delta_k(G)$ is not vertex decomposable. □

Proposition 4.3. *Let G be a double starlike tree such that is not a path. Then $\Delta_t(G)$ is vertex decomposable for all $2 \leq t \leq n+2$.*

Proof. Let $G = H(p,n,q)$ denote the double starlike tree obtained by attaching p pendant vertices to one pendant vertex of the path Pn and q pendant vertices to the other pendant vertex of Pn. We prove the theorem by induction on p the number of pendant vertices to pendant vertex x_1 of Pn. If $p = 0$ or $p = 1$, then by proposition 3.2 $\Delta_t(G)$ is vertex decomposable. If $p = 2$, then by Lemma 4.3 $\Delta_t(G)$ is vertex decomposable. Now, let $p > 2$ and $\{y_1, \ldots, y_p\}$ be p pendant vertices to pendant vertex x_1 of Pn. Since $G\backslash\{y_1\}$ is again double starlike tree on $p-1$ pendant vertices. Therefore, by induction hypothesis, $\Delta_t(G\backslash\{y_1\})$ is vertex decomposable. So $\Delta_t(G\backslash\{y_1\}) = \Delta_t(G)\backslash\{y_1\}$ is vertex decomposable. Let $t = 2$, then $\text{link}_{\Delta_2(G)}\{y_1\} = \langle\{x_1\}\rangle$ is simplex and vertex decomposable. Let $t = 3$, then $\text{link}_{\Delta_3(G)}\{y_1\} = \left\langle\{x_2,x_1\},\{y_2,x_1\}, \ldots, \{y_p,x_1\}\right\rangle$ is vertex decomposable. Let $3 < t \leq n+1$, then $\text{link}_{\Delta_t(G)}\{y_1\} = \langle\{x_1,x_2, \ldots, x_{t-1}\}\rangle$ is simplex and vertex decomposable. Let $t = n+2$, then $\text{link}_{\Delta_t(G)}\{y_1\} = \langle\{x_1, \ldots, x_n, y_1\}, \{x_1, \ldots, x_n, y_2\}, \ldots, \{x_1, \ldots, x_n, y_p\rangle$ is a path complex of a starlike tree which is vertex decomposable. It is easy to see that no face of $\text{link}_{\Delta_t(G)}\{y_1\}$ is a facet of $\Delta_t(G)\backslash\{y_1\}$. So $\Delta_t(G)$ is vertex decomposable.

Now, we are ready that prove the main result of this section.

Theorem 4.1. *Let G be a tree such that is not a path. Then $\Delta_t(G)$ is vertex decomposable for all $t \geq 2$ if and only if $G = H(p,n,q)$ or $H(p,n)$.*

Proof. $(\Rightarrow)$We prove by contradiction. Suppose $G \neq H(p,n,q)$ and $G \neq H(p,n)$. So there exists two paths of maximum length k in G which contain L common vertices such that one of these vertices is a leaf. Therefore, by proposition 4.2 $\Delta_k(G)$ is not vertex decomposable, which is a contradiction. ($\Leftarrow$) By proposition 4.1 and Proposition 4.3, the proof is completed.

5. Vertex decomposability path complexes of dynkin graphs

Dynkin diagrams first appeared [15] in the connection with classification of simple Lie groups. Among Dynkin diagrams a special role is played by the simply laced Dynkin diagrams A_n, D_n, E_6, E_7, and E_8. Dynkin diagrams are closely related to coxter graphs that appeared in geometry (see [16]).

After that, Dynkin diagrams appeared in many branches of mathematics and beyond, in particular in representation theory. In [17], Gabriel introduced a notion of a quiver (directed graph) and its representations. He proved the famous Gabriel's theorem on representation of quivers over algebraic closed field. Let Q be a finite quiver and $\overline{Q}$ the undirected graph obtained from Q by deleting the orientation of all arrows.

Theorem 5.1. *(Gabriel theorem). A connected quiver Q is of Finite type if and only if the graph $\overline{Q}$ is one of the following simply laced Dynkin diagrams:* A_n, D_n, E_6, E_7 *or* E_8.

Let L_n be a line graph on vertices $\{x_1, \dots, x_n\}$ and $\{x_j, y_j\}$ be whisker of L_n at x_j with $3 \le j \le n-1$.

We obtain some condition that $\Delta_t(L_n \cup W(x_j))$ is vertex decomposable, as an application of our results, vertex decomposability path complexes of Dynkin graphs are shown. Throughout this section, we assume $L_n \cup W(x_j)$ be an undirected graph. By Lemma 3.1, we have the following corollary:

Corollary 5.1. *Let A_n be a Dynkin graph and $2 \le t \le n$. Then $\Delta_t(A_n)$ is vertex decomposable.*

Proposition 5.1. *Let L_n be a line graph on vertices $\{x_1, \dots, x_n\}$ and $\{x_{n-1}, y_{n-1}\}$ be whisker of L_n at x_{n-1}.*

Then, $\Delta_t(L_n \cup W(x_{n-1}))$ is vertex decomposable for all $2 \le t \le n$.

Proof. Then by proposition 4.1 proof is trivial.

Corollary 5.2. *Let D_n be a Dynkin graph and $n \ge 4$. Then $\Delta_t(D_n)$ is vertex decomposable for all $2 \le t \le n$.*

Proof. We know that $D_n = L_n \cup W(x_{n-1})$. So by proposition 4.3 $\Delta_t(D_n)$ is vertex decomposable. □

Theorem 5.2. *Let L_n be a line graph on vertices $\{x_1, \dots, x_n\}$ and $\{x_j, y_j\}$ be whisker of L_n at x_j with $3 \le j \le n-2$.*

Then, $\Delta_t(L_n \cup W(x_j))$ is vertex decomposable if and only if $2 \le t \le 3$ or $n \ge t > \alpha$, where $\alpha = \min\{d(y_j, x_1), d(y_j, x_n)\}$.

Proof. We first show that $\Delta_t(L_n \cup W(x_j))$ is not vertex decomposable for all $4 \le t \le \alpha$. It is well known that if Δ is a Cohen-Macaulay simplicial complex, then $link_\Delta\{F\}$ is Cohen-Macaulay for each face F of Δ.

Also, we know that all Cohen-Macaulay complexes of positive dimension are connected. It is easy to see that $link_{\Delta_t(L_n \cup W(x_j))}\{x_j, y_j\}$ is disconnected and pure of positive dimension.

This implies that $\Delta_t(L_n \cup W(x_j))$ is not Cohen- Macaulay and hence $\Delta_t(L_n \cup W(x_j))$ is not vertex decomposable without loss of generality we can assume that $\alpha = d(y_j, x_1)$ if $t = 2$ or $n \ge t > \alpha$, one has:

$$\Delta_t(L_n \cup W(x_j)) = \langle \{x_1, \dots, x_t\}, \{x_2, \dots, x_{t+1}\}, \dots,$$
$$\{x_{j-1}, x_j, \dots, x_{j+t-2}\}, \{y_j, x_j, x_{j+1}, \dots, x_{j+t-2}\}, \dots, \{x_{n-t+1}, \dots, x_n\}\rangle$$

So $\Delta_t(L_n \cup W(x_j))\backslash y_j = \Delta_t(L_n)$ and by Lemma [4, 6, 7, 9, 10, 18–20] $\Delta_t(L_n \cup W(x_j))\backslash y_j$ is vertex decomposable. On the other hand, we have $link_{\Delta_t(L_n \cup W(x_j))}\{y_j\} = \langle \{x_j, x_{j+1}, \dots, x_{j+t-2}\}\rangle$ that is a simplex and vertex decomposable.

If $t = 3$, then

$$\Delta_3(L_n \cup W(x_j)) = \langle \{x_1, x_2, x_3\}, \{x_2, x_3, x_4\}, \dots,$$
$$\{x_{j-1}, x_j, y_j\}, \{y_j, x_j, x_{j+1}\}, \dots, \{x_{n-2}, x_{n-1}, x_n\}\rangle$$

and $\Delta_3(L_n \cup W(x_j))\backslash y_j = \Delta_3(L_n)$ which is vertex decomposable.

It is easy to see that $link_{\Delta_t(L_n \cup W(x_j))}\{y_j\} = \langle \{x_{j-1}, x_j\}, \{x_j, x_{j+1}\}\rangle$ is vertex decomposable and y_j is a shedding vertex.

Corollary 5.3. *Let E_6 be a Dynkin graph. Then $\Delta_t(E_6)$ is vertex decomposable if and only if $2 \leq t \leq 3$ or $t = 5$.*

Proof. Since $E_6 = L_5 \cup W(x_3)$, so by Theorem 5.2 the proof is completed.

Corollary 5.4. *Let E_7 be a Dynkin graph. Then $\Delta_t(E_7)$ is vertex decomposable if and only if $2 \leq t \leq 3$ or $5 \leq t \leq 6$.*

Proof. We know that $E_7 = L_6 \cup W(x_3)$ and the proof follow from Theorem 5.2.

Corollary 5.5. *Let E_8 be a Dynkin graph. Then $\Delta_t(E_8)$ is vertex decomposable if and only if $2 \leq t \leq 3$ or $5 \leq t \leq 7$.*

6. Stanley decompositions

Let R be any standard graded K-algebra over an infinite field K, *i.e*, R is a finitely generated graded algebra $R = \oplus_{i \geq 0} R_i$ such that $R_0 = K$ and R is generated by R_1. There are several characterizations of the depth of such an algebra. We use the one that depth(R) is the maximal length of a regular R-sequence consisting of linear forms. Let $x_F = \prod_{i \in F} x_i$ be a squarefree monomial for some $F \subseteq [n]$ and $Z \subseteq \{x_1, \dots, x_n\}$. The K-subspace $x_F K[Z]$ of $S = K[x_1, \dots, x_n]$ is the subspace generated by monomials $x_F u$, where u is a monomial in the polynomial ring $K[Z]$. It is called a squarefree Stanley space if $\{x_i : i \in F\} \subseteq Z$. The dimension of this Stanley space is $|Z|$. Let Δ be a simplicial complex on $\{x_1, \dots, x_n\}$. A squarefree Stanley decomposition $\mathcal{D}$ of $K[\Delta]$ is a finite direct sum $\oplus_i u_i K[Z_i]$ of squarefree Stanley spaces, which is isomorphic as a $\mathbb{Z}^n$-graded K-vector space to $K[\Delta]$, *i.e.*

$$K[\Delta] \cong \bigoplus_i u_i K[Z_i] \tag{21}$$

We denote by sdepth$(\mathcal{D})$ the minimal dimension of a Stanley space in $\mathcal{D}$ and we define sdepth$(K[\Delta]) = \max\{\text{sdepth}(\mathcal{D})\}$, where $\mathcal{D}$ is a Stanley decomposition of $K[\Delta]$. Stanley conjectured in [9] the upper bound for the depth of $K[\Delta]$ as the following:

$$\text{depth}(K[\Delta]) \leq \text{sdepth}(K[\Delta]) \tag{22}$$

Also, we recall another conjecture of Stanley. Let Δ be again a simplicial complex on $\{x_1, \dots, x_n\}$ with facets $G_1, \dots, G_t$. The complex Δ is called partitionable if there exists a partition $\Delta = \cup_{i=1}^{t}[F_i, G_i]$ where $F_i \subseteq G_i$ are suitable faces of Δ. Here, the interval $[F_i, G_i]$ is the set of faces $\{H \in \Delta : F_i \subseteq H \subseteq G_i\}$. In [10, 18] respectively Stanley conjectured each Cohen-Macaulay simplicial complex is partitionable [16–20]. This conjecture is a special case of the previous conjecture. Indeed, Herzog, Soleyman Jahan, and Yassemi [14] proved that for Cohen-Macaulay simplicial complex Δ on $\{x_1, \dots, x_n\}$ we have that $\text{depth}(K[\Delta]) \leq \text{sdepth}(K[\Delta])$ if and only if Δ is partitionable. Since each vertex decomposable simplicial complex is shellable and each shellable complex is partitionable [4, 6, 7, 19, 20]. Then as a consequence of our results, we obtain:

Corollary 6.1. *if $n = t$ or $t + 1$ then $\Delta_t(C_n)$ is partitionable and Stanley's conjecture holds for $K[\Delta_t(C_n)]$.*

Corollary 6.2. *Let G be a tree such that is not a path. if $G = H(p, n, q)$ or $G = H(p, n)$ then $\Delta_t(G)$ is vertex decomposable for all $t \geq 2$ and Stanley's conjecture holds for $K[\Delta_t(G)]$.*

Classification

2010 Mathematics Subject Classification, 13F20; 05E40; 13F55.

Author details

Seyed Mohammad Ajdani and Francisco Bulnes*

1 Department of Mathematics, Zanjan Branch, Islamic Azad University, Zanjan, Iran

2 IINAMEI, Investigación Internacional Avanzada en Matemáticas e Ingeniería, Chalco, Mexico

*Address all correspondence to: francisco.bulnes@tesch.edu.mx

References

[1] Björner A. Topological methods. In: Handbook of Combinatorics. Vol. 1, 2. Amsterdam: Elsevier; 1995. pp. 1819-1872

[2] Conca A, De Negri E. M-sequences, graph ideals and ladder ideals of linear type. Journal of Algebra. 1999;**211**(2): 599-624

[3] Francisco CA, Van Tuyl A. Sequentially Cohen-Macaulay edge ideals. Proceedings of the American Mathematical Society. 2007;**8**:2327-2337

[4] Saeedi Madani S, Kiani D, Terai N. Sequentially Cohen-Macaulay path ideals of cycles. Bull. Math. Soc. Sci. Math. Roumanie Tome. 2011;54**(102)** (4):353-363

[5] Aliloee A, Faridi S. Betti Numbers of the Path Ideals of Cycles and Lines, Preprint. 30 Oct 2011. arXiv: 1110.6653v1 [math.AC]

[6] Villareal RH. Cohen-Macaulay graphs. manuscripta mathematica. 1990;**66**(3):277-293

[7] Faridi S. The facet ideal of a simplicial complex. manuscripta mathematica. 2002;**109**:159-174

[8] He J, Van Tuyl A. Algebraic properties of the path ideal of a tree. Communications in Algebra. 2010;**38**: 1725-1742

[9] Stanley RP. Linear Diophantine equations and local cohomology. Inventiones Mathematicae. 1982;**68**(2): 175-193

[10] Stanley RP. Combinatorics and commutative algebra. In: Progress in Mathematics. Second ed. Vol. 41. Boston: Birkhauser; 1996

[11] Eagon JA, Reiner V. Resolutions of Stanley-Reisner rings and Alexander duality. Journal of Pure and Applied Algebra. 1998;**130**:265-275

[12] Herzog J, Hibi T. Monomial ideals. In: GTM 260. London: Springer; 2010

[13] Herzog J, Hibi T. Componentwise linear ideals. Nagoya Mathematical Journal. 1999;**153**:141-153

[14] Herzog J, Soleyman Jahan A, Yassemi S. *Stanley decompositions and partitionable simplicial complexes*. J. Alger. Comb. 2008;**27**:113-125

[15] Dynkin EB. Classification of the simple Lie groups. (Russian) Rec. Math, mat. Spornik (N.S.). 1946;**18**(60): 347-352

[16] Coxeter HSM. Regular poly to Pes. Third ed. New York: Dover Publications, Inc.; 1973

[17] Gabriel P. Unzertellungen I. Mansuscr. Math. 1972;**6**:71-103

[18] Stanley RP. Positivity problems and conjectures in algebraic combinatorics. In: Mathematics: Frontiers and Perspectives. Providence, RI: Amer. Math. Soc.; 2000. pp. 295-319

[19] Sampathkumar E, Reddy PSK, Subramanya MS. The line n- sigraph of a Symmetric n- sigraph. Southeast Asian Bulletin of Mathematics. 2010;**34**(5): 953-958

[20] Woodroofe R. Vertex decomposable graphs and obstructions to shellability. Proceedings of the American Mathematical Society. 2009;**137**: 3235-3246

Section 5

Special Compactness and Separability in Topological Spaces

Chapter 7

β_I-Compactness, β_I^*-Hyperconnectedness and β_I-Separatedness in Ideal Topological Spaces

Glaisa T. Catalan, Michael P. Baldado Jr and Roberto N. Padua

Abstract

Let (X, τ, I) be an ideal topological space. A subset A of X is said to be β-open if $A \subseteq \mathrm{cl}(\mathrm{int}(\mathrm{cl}(A)))$, and it is said to be β_I-open if there is a set $O\,(\in\tau)$ with the property (1) $O - A \in I$ and (2) $A - \mathrm{cl}(\mathrm{int}(\mathrm{cl}(O))) \in I$. The set A is called β_I-compact if every cover of A by β_I-open sets has a finite sub-cover. The set A is said to be $c\beta_I$-compact, if every cover $\{O_\lambda : \lambda \in \Lambda\}$ of A by β-open sets, Λ has a finite subset Λ_0 such that $A - \cup\{O_\lambda : \lambda \in \Lambda_0\} \in I$. The set A is said to be countably β_I-compact if every countable cover of A by β_I-open sets has a finite sub-cover. An ideal topological space (X, τ, I) is said to be β_I^*-hyperconnected if $X - \mathrm{cl}^*(A) \in I$ for every non-empty β_I-open subset A of X. Two subsets A and B of X is said to be β_I-separated if $\mathrm{cl}_{\beta_I}(A) \cap B = \varnothing = A \cap \mathrm{cl}_\beta(B)$. Moreover, A is called a β_I-connected set if it can't be written as a union of two β_I-separated subsets. An ideal topological space (X, τ, I) is called β_I-connected space if X is β_I-connected. In this article, we give some important properties of β_I-open sets, β_I-compact spaces, $c\beta_I$-compact spaces, β_I^*-hyperconnected spaces, and β_I-connected spaces.

Keywords: β-open set, β_I-open set, β_I-compact, $c\beta_I$-compact, $\beta^*{}_I$-hyperconnected

1. Introduction

Let (X, τ) be a topological space. A subset A of X is said to be a β-open set [1] if $A \subseteq \mathrm{cl}(\mathrm{int}(\mathrm{cl}(A)))$. For example, consider the topology $(X, \tau) = (\{a, b, c\}, \{\varnothing, \{a\}, \{a, b\}\{a, c\}, X\})$. Then $\varnothing$, X, $\{a, b\}$, $\{a, c\}$ are the β-open sets of (X, τ). A subset A of X is said to be *semi-open* set [1] if $A \subseteq \mathrm{cl}(\mathrm{int}(A))$. A subset A of X is said to be *α-open* set [2] if $A \subseteq \mathrm{int}(\mathrm{cl}(\mathrm{int}(A)))$. A subset A of X is said to be *pre-open* set [3] if $A \subseteq \mathrm{int}(\mathrm{cl}(A))$. A subset A of X is said to be *regular-open* set [4] if $A = \mathrm{int}(\mathrm{cl}(A))$. A subset A of X is said to be *β^*-open* set [5] if $A \subseteq \mathrm{cl}(\mathrm{int}(\mathrm{cl}(A))) \cup \mathrm{int}(\mathrm{cl}_\delta(A))$ (please see [5] for the notation $\mathrm{cl}_\delta(A)$). A subset A of X is said to be *$\hat{\beta}$-generalized-closed* set [6] if $\mathrm{cl}(\mathrm{int}(\mathrm{cl}(A))) \subseteq O$ whenever $A \subseteq O$ and O is open in X.

IntechOpen

An ideal I on a set X is a nonempty collection of subsets of X which satisfies the conditions: (1) $A \in I$ and $B \subseteq A$ implies $B \in I$, (2) $A \in I$ and $B \in I$ implies $A \cup B \in I$. Let (X, τ) be a topological space and I be an ideal in X. Then we call (X, τ, I) an ideal topological space. For example, let $X = \{a, b, c\}$. Then $I = \{\varnothing, \{a\}\}$ is an ideal on X. To see this, we note that the subsets of $\varnothing$ is itself, and the subsets of $\{a\}$ are $\{a\}$ and $\varnothing$. Note that all of these subsets are in I. Next, we observe that $\varnothing \cup \varnothing = \varnothing \in I$, $\varnothing \cup \{a\} = \{a\} \in I$ and $\{a\} \cup \{a\} = \{a\} \in I$. Thus, $I = \{\varnothing, \{a\}\}$ is an ideal on X.

For the concepts that were not discussed here please refer to [5, 7–13].

Topology is a new subject of mathematics, being born in the nineteenth century. However, the involvement of topology is clear in the other branches of math [12].

Topology is also seen in some fields of science. In particular, it is applied in biochemistry [14] and information systems [15].

Topology as a mathematical system is fundamentally comprised of open sets, among others. Open sets were generalized in a couple of different ways over the past. To mention a few, Stone [4] presented regular open set. Levine [1] presented semi-open sets. Najasted [2] presented α-open sets. Mashhour et al. [3] presented pre-open sets. Abd El-Monsef et al. [7] presented β-open set. Among these generalization, this study focused on one—the β-open sets.

Abd El-Monsef et al. [7] also presented the concepts β-continuous and β-open mappings. They gave some of their properties. Recently, β-open sets were investigated by many math enthusiast. For example, Abid [16] utilized β-open sets to gain some properties of non-semi-predense set. Tahiliani [13] presented an operation involving β-open sets which paved way to the creation of β-γ-open sets. Kannan and Nagaveni [6] generalized β-open set, and named it $\hat{\beta}$-generalized closed set. Mubarki et al. [5] also generalized β-open set, and named it β^*-open set. El-Mabhouh and Mizyed [17] also generalized β-open set, and named it βc-open set. Akdag and Ozkan [8] made an investigation of β-open sets in soft topological spaces. Arockiarani and Arokia Lancy [9] introduced $g\beta$-closed set and $gs\beta$-closed set (these were defined using β-open sets).

The notion of ideal topological spaces was introduced by Kuratowski [18]. Later, Vaidyanathaswamy [19] studied the concept in point set topology. Tripathy and Shravan [20, 21], Tripathy and Acharjee [22], Triapthy and Ray [23], among others, were also some of those who studied ideal topological spaces.

This study have important applications in some areas of mathematics. In particular, β_I-compactness, β_I^*-hyperconnectedness and β_I-separatedness can be investigated in the areas of measure theory, continuum theory and dimension theory just as the parallel notions (compactness, hypercompactness, and separatedness, respectively) were studied in those areas. The purpose of this paper is to introduce and study a notion of connectedness, hypercompactness, and separatedness relative to the family of all β-open sets in some ideal topological spaces.

2. β_I-compactness in ideal spaces

In this section, we gave some important properties of β_I-open sets in τ_ω-spaces.

Recall, a topological space (X, τ) is said to be a τ_ω-space if for every subset A of X, it is always true that $\text{int}(\text{cl}(A)) = \text{int}(\text{cl}(\text{int}A))$. For example, let $X = \{w, x, y, z\}$. Then $\tau_1 = \{\varnothing, \{w\}, \{w, x\}, \{w, y\}, \{w, x, y\}, X\}$ is a τ_ω-space, while $\tau_2 = \{\varnothing, \{w, x, y\}, X\}$ is not. Also, a discrete space is a τ_ω-space, while an indiscete space is not.

Lemma 1.1 characterizes β-open sets in a τ_ω-space.

Lemma 1.1. *Let (X,τ) be a τ_ω-space and I be an ideal in X. A set $A\ (\subseteq X)$ is a β-open set precisely if there is a set $O\ (\in\tau)$ with the property that $O\subseteq A\subseteq \mathrm{cl}(\mathrm{int}(\mathrm{cl}(O)))$.*

Proof: Suppose that A is a β-open set. Then $A\subseteq \mathrm{cl}(\mathrm{int}(\mathrm{cl}(A)))$. Consider $O=\mathrm{int}(A)$ (note that O is open). Since (X,τ) is a τ_ω-space, $\mathrm{int}(\mathrm{cl}(A))=\mathrm{int}(\mathrm{cl}(\mathrm{int}A))$. Hence, $O\subseteq A\subseteq \mathrm{cl}(\mathrm{int}(\mathrm{cl}(A)))=\mathrm{cl}(\mathrm{int}(\mathrm{cl}(\mathrm{int}(A))))=\mathrm{cl}(\mathrm{int}(\mathrm{cl}(O)))$.

Conversely, suppose that there is a set $O\ (\in\tau)$ with the property that $O\subseteq A\subseteq \mathrm{cl}(\mathrm{int}(\mathrm{cl}(O)))$. Since $O\subseteq A$, we have $\mathrm{cl}(O)\subseteq \mathrm{cl}(A)$. And so, $\mathrm{int}(\mathrm{cl}(O))\subseteq \mathrm{int}(\mathrm{cl}(A))$. Therefore, $\mathrm{cl}(\mathrm{int}(\mathrm{cl}(O)))\subseteq \mathrm{cl}(\mathrm{int}(\mathrm{cl}(A)))$ Thus, $A\subseteq \mathrm{cl}(\mathrm{int}(\mathrm{cl}(A)))$.

□

Next, we define β_I-open set.

Definition 1.1. *Let (X,τ) be a topological space and I be an ideal in X. A subset A of X is called β-open with respect to the ideal I, or a β_I-open set, if there exists an open set O such that (1) $O-A\in I$, and (2) $A-cl(int(cl(O)))\in I$.*

For example, let $X=\{a,b,c\}$, $\tau=\{\varnothing,\{a\},\{b,c\},X\}$, and $I=\{\varnothing,\{b\}\}$ (note that τ is a topology on X, and I is an ideal on X). Then $A=\{b,c\}$ is a β_I-open set. To see this, consider $O=\{b,c\}$. Then O is a open set. Observe that $O-A=\{b,c\}-\{b,c\}=\varnothing\in I$, and $A-\mathrm{cl}(\mathrm{int}(\mathrm{cl}(O)))=\{b,c\}-\mathrm{cl}(\mathrm{int}(\mathrm{cl}(\{b,c\})))=\{b,c\}-\mathrm{cl}(\mathrm{int}(\{b,c\}))=\{b,c\}-\mathrm{cl}(\{b,c\})=\{b,c\}-\{b,c\}=\varnothing\in I$. Thus, $A=\{b,c\}$ is β-open with respect to the ideal I.

Lemma 1.2 says that an open set is a β_I-open set, and an element of the ideal is a β_I-open set. One may see [24] to gain more insights relative to these ideas. While, Lemma 1.3 says that in a τ_ω-space a β-open set is also a β_I-open set.

Lemma 1.2. *Let (X,τ) be a topological space and I be an ideal in X. Then the following statements are true.*

i. Every open set is a β_I-open set.

ii. Every element of I is a β_I-open set.

Proof: (i) Let A be an open set. Note that $A-A=\varnothing\in I$, and $A-\mathrm{cl}(\mathrm{int}(\mathrm{cl}(A)))\subseteq A-\mathrm{cl}(A)=\varnothing\in I$. Thus, A is β_I-open. (ii) Let $A\in I$. Consider $O=\varnothing$. Note that $O-A=\varnothing-A=\varnothing\in I$, and $A-\mathrm{cl}(\mathrm{int}(\mathrm{cl}(O)))=A-\varnothing=A\in I$. Thus, A is β_I-open. □

Lemma 1.3. *Let (X,τ) be a τ_ω-space and I be an ideal in X. Then every β-open set is a β_I-open set.*

Proof: Let A be a β-open set. By Lemma 1.1 there exists an open set O such that $O\subseteq A\subseteq \mathrm{cl}(\mathrm{int}(\mathrm{cl}(O)))$. Hence $O-A=\varnothing\in I$, and $A-\mathrm{cl}(\mathrm{int}(\mathrm{cl}(O)))=\varnothing\in I$. Thus, A is β_I-open. □

Let (X,τ) be a topology and I be an ideal in X. We say that I is countably additive if $\cup\{A_i: i\in\mathbb{N}\}\in I$ whenever $\{A_i: i\in\mathbb{N}\}$ is a (countable) family of elements of I.

Lemma 1.4 says that in a τ_ω-space, if I is the minimal ideal, then the β_I-open sets are precisely the β-open sets.

Lemma 1.4. *Let (X,τ) be a τ_ω-space and I be an ideal in X. If I is not countably additive, then the following statements are equivalent.*

i. $I=\{\varnothing\}$.

ii. *A is a β-open set precisely when A is a β_I-open set.*

Proof: $(i)\Rightarrow(ii)$ Suppose that $I=\{\varnothing\}$. Let A be a β-open set. By Lemma 1.3, A is a β_I-open set. For the converse, let A be a β_I-open set and O be an open set with $O-A\in I$ and $A-\mathrm{cl}(\mathrm{int}(\mathrm{cl}(O)))\in I$. Because $I=\{\varnothing\}$, we have $O-A=\varnothing$ and $A-\mathrm{cl}(\mathrm{int}(\mathrm{cl}(O)))=\varnothing$. Hence, $O\subseteq A$ and $A\subseteq \mathrm{cl}(\mathrm{int}(\mathrm{cl}(O)))$, that is $O\subseteq A\subseteq \mathrm{cl}(\mathrm{int}(\mathrm{cl}(O)))$. Therefore, by Lemma 1.1 A is β-open.

$(ii) \Rightarrow (i)$ Suppose that (ii) holds, and that $I \neq \{\varnothing\}$. Let D be a non-empty element of I. By Lemma 1.2, D is β_I-open. Thus, by assumption D is β-open. Now, by Lemma 1.1, there exists $O_1 \in \tau$ with $O_1 \subseteq D \subseteq \mathrm{cl}(\mathrm{int}(\mathrm{cl}(O_1)))$. Since D is an element of I and $O_1 \subseteq D$, we have $O_1 \in I$. Hence, $O_1 \cup D \in I$. By Lemma 1.1, $O_1 \cup D$ is a β_I-open set. Hence, by assumption $O_1 \cup D$ is a β-open set. And so, again there exists $O_2 \in \tau$ with $O_2 \subseteq (O_1 \cup D) \subseteq \mathrm{cl}(\mathrm{int}(\mathrm{cl}(O_2)))$. Since $O_1 \cup D \in I$ and $O_2 \subseteq O_1 \cup D$, we have $O_2 \in I$. Hence, $O_1 \cup O_2 \cup D \in I$. Thus, by Lemma 1.1, $O_1 \cup O_2 \cup D$ is a β_I-open set. By assumption $O_1 \cup O_2 \cup D$ is a β-open set. And so, again there exists $O_3 \in \tau$ with $O_3 \subseteq (O_1 \cup O_2 \cup D) \subseteq \mathrm{cl}(\mathrm{int}(\mathrm{cl}(O_3)))$. Since $O_1 \cup O_2 \cup D \in I$ and $O_3 \subseteq O_1 \cup O_2 \cup D$, we have $O_3 \in I$. Hence, $O_1 \cup O_2 \cup O_3 \cup D \in I$. Continuing in this fashion we obtain a countably infinite subset $\{O_1, O_2, O_3, \dots\}$ of I with $O_1 \cup O_2 \cup O_3 \cup \cdots \in I$. This is a contradiction since I is not countably additive. Thus, $I = \{\varnothing\}$. □

Next, we define β_I-compact set, β_I-compact space, compatible β_I-compact set, and compatible β_I-compact space.

Definition 1.2. *Let (X, τ, I) be an ideal topological space. A subset A of X is said to be β_I-compact if for every cover $\{O_\lambda : \lambda \in \Lambda\}$ of A by β_I-open sets, Λ has a finite subset Λ_0, such that $\{O_\lambda : \lambda \in \Lambda_0\}$ still covers A. A space X is said to be a β_I-compact space if it is β_I-compact as a subset.*

Definition 1.3. *Let (X, τ, I) be an ideal topological space. A subset A of X is said to be countably β_I-compact if for every countable cover $\{O_n : n \in \mathbb{N}\}$ of A by β_I-open sets, $\mathbb{N}$ has a finite subset $\{i_j : j = 1, 2, \dots, k\}$ with the property that $\left\{O_{i_j} : j = 1, 2, \dots, k\right\}$ still covers A. A space X is said to be a countably β_I-compact space if it is countably β_I-compact as a subset.*

Definition 1.4. *Let (X, τ, I) be an ideal topological space. A subset A of X is said to be compatible β_I-compact, or simply $c\beta_I$-compact, if for every cover $\{O_\lambda : \lambda \in \Lambda\}$ of A by β-open sets, Λ has a finite subset Λ_0, such that $A - \cup\{O_\lambda : \lambda \in \Lambda_0\} \in I$. An ideal topological space (X, τ, I) is said to be $c\beta_I$-compact space if it is $c\beta_I$-compact as a subset.*

Theorem 1.1 says that in an ideal τ_ω-space in which I is the minimal ideal, the notions β-compact, β_I-compact and $c\beta_I$-compact coincides.

Theorem 1.1. *Let (X, τ) be a τ_ω-space and $I = \{\varnothing\}$. Then the following statements are equivalent.*

i. *(X, τ, I) is a β-compact space.*

ii. *(X, τ, I) is a β_I*-compact space.

iii. *(X, τ, I) is a $c\beta_I$-compact space.*

Proof: $(i) \Rightarrow (ii)$ Suppose that (i) holds. Let $\{O_\lambda : \lambda \in \Lambda\}$ be a family of β-open sets that covers X. By assumption, Λ has a finite subset, say Λ_0, with the property that $\{O_\lambda : \lambda \in \Lambda_0\}$ still covers X. By Lemma 1.3 (iii), $\{O_\lambda : \lambda \in \Lambda_0\}$ is also a family of β_I-open sets. Hence, $\{O_\lambda : \lambda \in \Lambda_0\}$ is a finite covering of X by β_I-open sets. Therefore, X is β_I compact.

$(ii) \Rightarrow (iii)$ Suppose that (ii) holds. Let $\{O_\lambda : \lambda \in \Lambda\}$ be a family of β-open sets that covers X. Since $I = \{\varnothing\}$, by Lemma 1.4 $\{O_\lambda : \lambda \in \Lambda\}$ is also a family of β_I-open sets that covers X. By assumption, Λ has a finite subset, say Λ_0, with the property that $\{O_\lambda : \lambda \in \Lambda_0\}$ still covers X. Note that $\{U_\lambda : \lambda \in \Lambda_0\}$ is also a family of β-open sets, and $X - \cup_{\lambda \in \Lambda_0} O_\lambda = \varnothing \in I$. Therefore, X is $c\beta_I$ compact.

$(iii) \Rightarrow (i)$ Suppose that (iii) holds. Let $\{O_\lambda : \lambda \in \Lambda\}$ be a family of β-open sets that covers X. By assumption, Λ has a finite subset, say Λ_0, with the property that

DOI: http://dx.doi.org/10.5772/intechopen.101524

$X - \cup_{\lambda \in \Lambda_0} O_\lambda \in I$. Since $I = \{\varnothing\}$, $X - \cup_{\lambda \in \Lambda_0} O_\lambda = \varnothing$, that is $X \subseteq \cup_{\lambda \in \Lambda_0} O_\lambda$. Hence, $\{O_\lambda : \lambda \in \Lambda_0\}$ is covering of X. Therefore, X is β compact. □

Theorem 1.2 presents a characterization of β_I-compact spaces.

Theorem 1.2. *Let (X, τ, I) be an ideal topological space. Then the following are equivalent.*

i. *(X, τ, I) is β_I-compact.*

ii. *If $\{F_\lambda : \lambda \in \Lambda\}$ is a family of β_I-closed sets with $\cap\{F_\lambda : \lambda \in \Lambda\} = \varnothing$, then Λ has a finite subset, say Λ_0, with the property $\cap\{F_\lambda : \lambda \in \Lambda_0\} = \varnothing$.*

Proof: $(i) \Rightarrow (ii)$ Suppose that (i) holds. Let $\{F_\lambda : \lambda \in \Lambda\}$ be a family of β_I-closed sets with the property $\cap\{F_\lambda : \lambda \in \Lambda\} = \varnothing$. Then $\cup\left\{F_\lambda^C : \lambda \in \Lambda\right\} = (\cap\{F_\lambda : \lambda \in \Lambda\})^C = X$. Hence, $\left\{F_\lambda^C : \lambda \in \Lambda\right\}$ is a family of β_I-open sets which covers of X. By assumption, Λ has a finite subset, say Λ_0, with the property $\cup\left\{F_\lambda^C : \lambda \in \Lambda_0\right\} = X$, i.e. $\cap\{F_\lambda : \lambda \in \Lambda_0\} = \varnothing$.

$(ii) \Rightarrow (i)$ Suppose that (ii) holds. Let $\{O_\lambda : \lambda \in \Lambda\}$ be a family of β_I-open sets that covers X, i.e. $\cup\{U_\lambda : \lambda \in \Lambda\} = X$. Then $\cap\left\{O_\lambda^C : \lambda \in \Lambda\right\} = (\cup\{U_\lambda : \lambda \in \Lambda\})^C = \varnothing$. Note that O^C is β_I-closed since O is β_I-open. By assumption, Λ has a finite subset, say Λ_0, with the property that $\cap\left\{O_\lambda^C : \lambda \in \Lambda_0\right\} = \varnothing$. Thus, $\cup\{O_\lambda : \lambda \in \Lambda_0\} = \left(\cap\left\{O_\lambda^C : \lambda \in \Lambda_0\right\}\right)^C = X$, that is $\{O_\lambda : \lambda \in \Lambda_0\}$ is a family of β_I-open sets that covers X.

□

Theorem 1.3 presents a characterization of β_I-compact spaces.

Theorem 1.3. *Let (X, τ) be a topological space and I be an ideal in X. Then the following are equivalent.*

i. *(X, τ, I) is $c\beta_I$-compact.*

ii. *If $\{F_\lambda : \lambda \in \Lambda\}$ is a family of β-closed sets with $\cap\{F_\lambda : \lambda \in \Lambda\} = \varnothing$, then Λ has a finite subset, say Λ_0, with the property that $\cap\{F_\lambda : \lambda \in \Lambda_0\} \in I$.*

Proof: $(i) \Rightarrow (ii)$ Suppose that (i) holds. Let $\{F_\lambda : \lambda \in \Lambda\}$ be a family of β-closed sets such that $\cap\{F_\lambda : \lambda \in \Lambda\} = \varnothing$. Note that $\cup\left\{F_\lambda^C : \lambda \in \Lambda\right\} = (\cap\{F_\lambda : \lambda \in \Lambda\})^C = X$. Hence, $\left\{F_\lambda^C : \lambda \in \Lambda\right\}$ is a family of β-open sets that covers X. By assumption, Λ has a finite subset, say Λ_0, with the property $X - \cup\left\{F_\lambda^C : \lambda \in \Lambda_0\right\} \in I$, i.e. $\cap\{F_\lambda : \lambda \in \Lambda_0\} \in I$.

$(ii) \Rightarrow (i)$ Suppose that (ii) holds. Let $\{O_\lambda : \lambda \in \Lambda\}$ be a family of β-open sets that covers X, i.e. $\cup\{O_\lambda : \lambda \in \Lambda\} = X$. Note that $\cap\left\{O_\lambda^C : \lambda \in \Lambda\right\} = (\cup\{O_\lambda : \lambda \in \Lambda\})^C = \varnothing$. By assumption, Λ has a finite subset, say Λ_0, with the property $\cap\left\{O_\lambda^C : \lambda \in \Lambda_0\right\} \in I$, i.e. $X - \cup\{O_\lambda : \lambda \in \Lambda_0\} \in I$.

□

Remark 1.1. *[25] Let (X, τ, I) and (Y, σ, J) be ideal spaces, and $f : X \to Y$ be a fuction. Then:*

i. $f(I) = \{f(A) : A \in I\}$ is an ideal in Y, where $f(A) = \{f(a) : a \in A\}$; And,

ii. if f is bijective, then $f^{-1}(J) = \left\{f^{-1}(B) : B \in J\right\}$ is an ideal in X, where $f^{-1}(B) = \left\{f^{-1}(b) : b \in B\right\}$.

Next, we define β_I-open, β_I-irresolute, and β_I-continuous functions.

Definition 1.5. *Let (X, τ, I) and (Y, σ, J) be ideal topological spaces. A function $f : X \to Y$ is said to be*

i. *β-open if $f(A)$ is β-open for every β-open set A,*

ii. *β-irresolute if $f^{-1}(B)$ is β-open for every β-open set B, and*

iii. *β-continuous if $f^{-1}(B)$ is β-open for every open set B.*

iv. *β_I-open if $f(A)$ is β_J-open for every β_I-open set A,*

v. *β_I-irresolute if $f^{-1}(B)$ is β_I-open for every β_J-open set B, and*

vi. *β_I-continuous if $f^{-1}(B)$ is β_I-open for every open set B.*

Theorem 1.4 says that given a β-irresolute function, if the domain is compatibly compact, then so is the image of f. On the other hand, Theorem 1.5 say that given an open surjection, if the co-domain is compatibly compact, then so is the domain.

Theorem 1.4. *Let (X, τ) and (Y, σ) be topological spaces, I be an ideal in X, and $f : X \to Y$ be a β-irresolute function. If X is a $c\beta_I$-compact space, then $(f(X), \{f(X) \cap B : B \in \sigma\}, f(I))$ is a $c\beta_{f(I)}$-compact space.*

Proof: Let $\{O_\lambda : \lambda \in \Lambda\}$ be a family of β-open sets that covers $f(X)$. Beacuse f is a β-irresolute, $\{f^{-1}(O_\lambda) : \lambda \in \Lambda\}$ is a family of by β-open sets that covers X. Because X is $c\beta_I$-compact, Λ has a finite subset, say Λ_0, with the property $X - \cup\{f^{-1}(O_\lambda) : \lambda \in \Lambda_0\} \in I$. Hence, by Remark 1.1 $f(X) - \cup\{O_\lambda : \lambda \in \Lambda_0\} = f(X - \cup\{f^{-1}(O_\lambda) : \lambda \in \Lambda_0\}) \in f(I)$. □

Theorem 1.5. *Let (X, τ) and (Y, σ) be topological spaces, J be an ideal in Y, and $f : X \to Y$ be a β-open surjection (surjective function). If Y is $c\beta_J$-compact, then X is $c\beta_{f^{-1}(J)}$-compact.*

Proof: Let $\{O_\lambda : \lambda \in \Lambda\}$ be a family of β-open sets that covers X. Beacuse f is a β-open surjection, $\{f(O_\lambda) : \lambda \in \Lambda\}$ is a family β-open sets that covers Y. Because Y is $c\beta_J$-compact, Λ has a finite subset, say Λ_0, with the property $Y - \cup\{f(O_\lambda) : \lambda \in \Lambda_0\} \in J$. Hence, $X - \cup\{O_\lambda : \lambda \in \Lambda_0\} = f^{-1}(Y - \cup\{f(O_\lambda) : \lambda \in \Lambda_0\}) \in f^{-1}(J)$. □

The next theorem says that in a τ_ω-space and when $I = \{\varnothing\}$, the family of all countably β_I-compact space contains all $c\beta_I$-compact space.

Theorem 1.6. *Let (X, τ, I) be an ideal τ_ω-space and $I = \varnothing$. If X is $c\beta_I$-compact, then it is also countably β_I-compact.*

Proof: Let $\{O_n : n \in \mathbb{N}\}$ be a countable family β_I-open sets that covers X. Because X is $c\beta_I$-compact, $\mathbb{N}$ has a finite subset $\{i_j : j = 1, 2, \ldots, k\}$ with the property that $X - \cup\left\{O_{i_j} : j = 1, 2, \ldots, k\right\} \in I$. Because $I = \{\varnothing\}$, $X = \cup\left\{O_{i_j} : j = 1, 2, \ldots, k\right\} \in I$. By Lemma 1.4 $\left\{O_{i_j} : j = 1, 2, \ldots, k\right\}$ is also a family of β-open sets. Hence, $\left\{O_{i_j} : j = 1, 2, \ldots, k\right\}$ is a finite subcover of X by β-open sets. □

3. β_I^*-hyperconnectedness in ideal spaces

The concept $*$-hyperconnectedness was introduced by Ekici et al. [26], and the concept $I*$-hyperconnectedness was introduced by Abd El-Monsef et al. [27].

These insights motivated us to create the concept called β_I^*-hyperconnectedness. One may see [28] to gain more insights on these ideas.

Definition 1.6. *Let (X, τ) be a topological space and I be an ideal on X. A function $(\ )^*(I, \tau) : P(X) \to P(X)$ given by $A^*(I, \tau) = \{x \in X : A \cup U \notin I \text{ for every } U \in \tau(x)\}$ where $\tau(x) = \{U \in \tau : x \in U\}$ is called a local of A with respect to τ and I.*

Example 1.1. *Let $X = \{a, b, c\}$, $\tau = \{\emptyset, \{a\}, \{b\}, \{c\}, \{a, b\}, \{a, c\}, \{b, c\}, X\}$, and $I = \{\emptyset, \{a\}, \{b\}, \{a, b\}\}$ (note that τ is a topology on X and I is an ideal on X). Then, $\emptyset^* = \emptyset$, $\{a\}^* = \{c\}$, $\{b\}^* = \{c\}$, $\{c\}^* = X$, $\{a, b\}^* = \{c\}$, $\{a, c\}^* = X$, $\{b, c\}^* = X$ and $X^* = X$.*

Definition 1.7. *Let (X, τ) be a topological space and I be an ideal on X. The Kuratowski closure operator $Cl()^*(I, \tau) : P(X) \to P(X)$ for the topology $\tau^*(I, \tau)$ is given by $Cl(A)^*(I, \tau) = A \cup A^*$.*

Example 1.2. *Consider the ideal space of Example 3. Then we have, $Cl(\emptyset)^* = \emptyset \cup \emptyset^* = \emptyset \cup \emptyset = \emptyset$, $Cl(\{a\})^* = \{a\} \cup \{a\}^* = \{a\} \cup \{c\} = \{a, c\}$, $Cl(\{b\})^* = \{b\} \cup \{b\}^* = \{b\} \cup \{c\} = \{b, c\}$, $Cl(\{c\})^* = \{c\} \cup \{c\}^* = \{c\} \cup X = X$, $Cl(\{a, b\})^* = \{a, b\} \cup \{a, b\}^* = \{a, b\} \cup \{c\} = X$, $Cl(\{a, c\})^* = \{a, c\} \cup \{a, c\}^* = \{a, c\} \cup X = X$, $Cl(\{b, c\})^* = \{b, c\} \cup \{b, c\}^* = \{b, c\} \cup X = X$, and $Cl(X)^* = X \cup X^* = X \cup X = X$.*

Definition 1.8. *Let (X, τ) be a topological space and I be an ideal on X. The Kuratowski interior operator $Int()^*(I, \tau) : P(X) \to P(X)$ for the topology $\tau^*(I, \tau)$ is given by $Int(A)^*(I, \tau) = X - Cl(X - A)^*$.*

Definition 1.9 is taken from [26], while Definition 1.10 is taken from [29].

Definition 1.9. *[26] An ideal space (X, τ, I) is called $*$-hyperconnected if $cl^*(A) = X$ for all non-empty open set $A \subseteq X$.*

Definition 1.10. *[29] An ideal space (X, τ, I) is called I^*-hyperconnected if $X - cl^*(A) \in I$ for all non-empty open set $A \subseteq X$.*

A notion similar to Definition 1.9 and Definition 1.10 is presented next.

Definition 1.11. *An ideal topological space (X, τ, I) is said to be β_I^*-hyperconnected space if $X - \mathrm{cl}^*(A) \in I$ for every non-empty β_I-open subset A of X.*

The next theorem says that the family of all $\beta_I *$-hyperconnected space contains all $I *$-hyperconnected space.

Theorem 1.7. *Let (X, τ) be a topological space, and I be an ideal in X. If X is $I *$-hyperconnected, then it is β_I^*-hyperconnected also.*

Proof: Let X be $I *$-hyperconnected, and A be a non-empty open set. Because X is I_*-hyperconnected, we have $X - cl(A)^* \in I$ for all non-empty open set $A \subseteq X$. And, because an open set is also a β_I-open set, we have $X - cl(A)^* \in I$ for all non-empty β_I-open set $A \subseteq X$. Hence, X is β_I^*-hyperconnected. □

The next lemma is clear.

Lemma 1.5. *Let (X, τ) be a topological space. Then the intersection of any family of ideals on X is an ideal on X.*

Theorem 1.8 is taken from [29]. It says that when I is the minimal ideal, then the notions $*$-hyperconnected and I^*-hyperconnected are equivalent.

Theorem 1.8. *[29] Let (X, τ) be a topological space, and $I = \{\emptyset\}$. Then, X is $*$-hyperconnected if and only if it is I^*-hyperconnected.*

The next remark is clear.

Remark 1.2. *If (X, τ) is a clopen topological space (a space in which every open set is also closed), then A is open if and only if A is β-open.*

To see this, let A be an open set. Since τ is clopen, A is closed also. Hence, $\mathrm{cl}(\mathrm{int}(\mathrm{cl}(A))) = A$. Thus, A is a β-open set. Conversely, if A is a β-open set, then $A \subseteq \mathrm{cl}(\mathrm{int}(\mathrm{cl}(A)))$. This implies that A must be open.

Theorem 1.9 says that in a clopen τ_ω-space, with respect to the minimal ideal I, the notions β_I^*-hyperconnected and I^*-hyperconnected are equivalent.

Theorem 1.9. *Let (X, τ) be a clopen τ_ω-space, and $I = \{\varnothing\}$. Then, X is I^*-hyperconnected if and only if it is β_I^*-hyperconnected.*

Proof: Suppose that X is I^*-hyperconnected. Let A is a non-empty element of τ. Then $X - cl^*(A) \in I$. By Remark 1.2 and Lemma 1.5, every open set is absolutely β_I-open. Thus, $X - cl^*(A) \in I$ for all β_I-open set A $(\neq \varnothing)$. Therefore, X is β_I^*-hyperconnected also. Conversely, suppose that X is β_I^*-hyperconnected. Let A be a non-empty β_I-open set. Then $X - cl^*(A) \in I$. By Remark 1.2 and Lemma 1.5, every β_I-open set is absolutely open. Thus, $X - cl(A)^* \in I$ for all open set A $(\neq \varnothing)$. Therefore, X is I^*-hyperconnected also. □

Corollary 1.1 says that in a clopen τ_ω-space, relative to the minimal ideal I, the notions β_I^*-hyperconnected, I^*-hyperconnected, and $*$-hyperconnected are equivalent.

Corollary 1.1. *Let (X, τ) be a clopen τ_ω-space and $I = \{\varnothing\}$. Then the following statements are equivalent.*

i. *X is I^*-hyperconnected.*

ii. *X is β_I^*-hyperconnected.*

iii. *X is β_I^*-hyperconnected.*

Theorem 1.3 may be an important property.

Remark 1.3. *If an ideal τ_ω space $(X, \tau, \{\varnothing\})$ is a $\beta_I *$-hyperconnected space, then $X - \mathrm{cl}^*(A) \in I$ for every non-empty β-open subset A of X.*

To see this, let A is a non-empty β-open set. By Lemma 1.4 A is β_I-open. Since X is β_I^*-hyperconnected, $X - cl^*(A) \in I$.

Theorem 1.10 is a characterization of β_I^*-hyperconnected space.

Theorem 1.10. *Let (X, τ) be an topological space and I be an ideal in X. Then the following statements are equivalent.*

i. *X is a β_I^*-hyperconnected space.*

ii. $\mathrm{Int}(A)^* \in I$ *for all proper β_I-closed subset A of X.*

Proof: $(i) \Rightarrow (ii)$ Suppose that (i) holds. Let B be β_I-closed. Then $X - B$ is β_I-open. Since $B \neq X, X - B \neq \varnothing$. Hence, by assumption we have $\mathrm{Int}(B)^* = X - \mathrm{Cl}(X - B)^* \in I$.

$(ii) \Rightarrow (i)$ Suppose that (ii) holds. Let A $(\neq X)$ be a non-empty β_I-open set. Then $X - A$ is a non-empty β_I-open set. Hence, by assumption we have $X - \mathrm{cl}(A)^* = X - \mathrm{cl}(X - (X - A))^* = \mathrm{int}(X - A)^* \in I$. Thus, X is β_I^*-hyperconnected. □

4. β_I-separatedness in ideal spaces

In this section, we present the concepts β_I-separated sets and β_I-connected sets. We also present some of their important properties.

Let (X, τ, I) be an ideal topological space and A be a subset of X. The β-closure of A, denoted by $\mathrm{cl}_\beta(A)$, is the smallest β-closed set that contains A. The β_I-closure of A, denoted by $\mathrm{cl}_{\beta_I}(A)$, is the smallest β_I-closed set that contains A.

Next, we define β_I-separated set, β_I-connected set, and β_I-connected space.

Definition 1.12. *Let (X, τ, I) be an ideal topological space. A pair of subsets, say A and B, of X is said to be β_I-separated if $cl_{\beta_I}(A) \cap B = \varnothing = A \cap cl_\beta(B)$.*

Definition 1.13. *Let (X, τ, I) be an ideal topological space and A be a subset of X. Then A is said to be β_I-connected if it cannot be expressed as a union of two β_I-separated sets. A topological space X is said to be β_I-connected if it is β_I-connected as a subset.*

Recall, a topological space (X, τ) is said to be a τ_ζ-space if for every pair of subsets A and B of X, it is always true that $\mathrm{cl}(A \cap B) = \mathrm{cl}(A) \cap \mathrm{cl}(B)$ and $\mathrm{int}(A \cap B) = \mathrm{int}(A) \cap \mathrm{int}(B)$. For example, a discrete space is a τ_ζ-space, while an indiscete space is not. Also, if $X = \{a, b, c\}$, then $\tau = \{\varnothing, \{a\}, X\}$ is not a τ_ζ-space. Let (X, τ) be a τ_ζ-space and I be an ideal in X. Then we call (X, τ, I) an ideal τ_ζ-space.

Lemma 1.6 present sufficient conditions for two sets to be β_I-separated.

Lemma 1.6. *Let (X, τ) be a topological space and I be an ideal in X. If A $(\neq \varnothing)$ is β-open and B $(\neq \varnothing)$ is β_I-open with $A \cap B = \varnothing$, then they are β_I-separated.*

Proof: Suppose that A and B is not β_I-separated, that is $\mathrm{cl}_{\beta_I}(A) \cap B \neq \varnothing$ or $A \cap \mathrm{cl}_\beta(B) \neq \varnothing$. Because $A \cap B = \varnothing$, we have $A \subseteq B^C$ and $B \subseteq A^C$. If A is β-open, then A^C is β-closed. Similarly, if B is β_I-open, then B^C is β_I-closed. Thus, $B^C \cap B \supseteq \mathrm{cl}_{\beta_I}(A) \cap B \neq \varnothing$, or $A \cap A^C \supseteq A \cap \mathrm{cl}_\beta(B) \neq \varnothing$. A contradiction. □

Lemma 1.7 says that in a τ_ω-space every β_I-connected space is connected. Recall, a space is connected if it cannot be written as a union of two non-empty open sets.

Lemma 1.7. *Let (X, τ) be a topology and I be an ideal in X. If X is β_I-connected, then it is connected.*

Proof: Suppose that to the contrary X is not connected. Let A and B be non-empty disjoint elements of τ with $X = A \cup B$. Note that A and B are β-open and β_I-open also. Because $A = B^C$ and $B = A^C$, A and B are also β-closed and β_I-closed. And so, $A = \mathrm{cl}_{\beta_I}(A)$ and $B = \mathrm{cl}_\beta(B)$. Thus, $\mathrm{cl}_{\beta_I}(A) \cap B = A \cap B = \varnothing$ and $A \cap \mathrm{cl}_\beta(B) = A \cap B = \varnothing$. This implies that X is β_I-separated, that is X is not β_I-connected. □

Remark 1.4. *Let (X, τ) be a topology and I be an ideal in X. If $Y \subseteq X$, then $I_Y = \{Y \cap A : A \in I\}$ is an ideal in the relative topology (Y, τ_Y).*

To see this, for the first property, let $B \in I_Y$ and $A \subseteq B$. Then $A \subseteq B \subseteq Y$. Now, if $A \in I_Y$, then there exist $C \in I$ such that $Y \cap C = A$. Note that $A \subseteq B \subseteq C$. Hence, $A, B \in I$. Thus, $A = Y \cap A \in I_Y$. Next, for the second, let $D, E \in I_Y$. Then $D \subseteq Y$ and $E \subseteq Y$. if $D \in I_Y$, then there exist $F \in I$ such that $Y \cap F = D$. Similarly, if $E \in I_Y$, then there exist $G \in I$ such that $Y \cap G = E$. Since I is an ideal, $F \cup G \in I$. Now, because $D \cup E \subseteq F \cup G$, $D \cup E \in I$. Thus, $D \cup E = (D \cup E) \cap Y \in I_Y$.

The next statement, Theorem 1.11, presents a way to construct β_I-open sets in a subspace.

Theorem 1.11. *Let (X, τ, I) be an ideal τ_ζ-space and Y be a clopen (a set that is open and closed at the same time) set. If A is a β_I-open subset of X (τ, I), then $A \cap Y$ is a β_{I_Y}-open set in Y (τ_Y, I_Y).*

Proof: Let A be a β_I-open set in X (τ, I). Then there exists an open set U' such that $U' - A \in I$ and $A - \mathrm{cl}(\mathrm{int}(\mathrm{cl}(U'))) \in I$. Let $U = U' \cap Y$. Then

$$\begin{aligned} U - (A \cap Y) &= U \cap (A \cap Y)^C \\ &= (U' \cap Y) \cap \left(A^C \cup Y^C\right) \\ &= \left(U' \cap Y \cap A^C\right) \cup \left(U' \cap Y \cap Y^C\right) \\ &= U' \cap Y \cap A^C \\ &= (U' - A) \cap Y \in I_Y. \end{aligned} \tag{1}$$

Moreover, since X is a τ_ζ-space and Y is clopen

$$\begin{aligned} (A \cap Y) - \mathrm{cl}(\mathrm{int}(\mathrm{cl}(U))) &= (A \cap Y) - \mathrm{cl}(\mathrm{int}(\mathrm{cl}(U' \cap Y))) \\ &= (A \cap Y) - \mathrm{cl}(\mathrm{int}(\mathrm{cl}(U'))) \cap Y \\ &= [A - \mathrm{cl}(\mathrm{int}(\mathrm{cl}(U')))] \cap Y \in I_Y. \end{aligned} \tag{2}$$

This shows that $A\cap Y$ is β_{I_Y}-open in $Y(\tau_Y, I_Y)$. □

Corollary 1.2. *Let (X,τ,I) be an ideal τ_ζ-space and Y be a clopen set. If A is a β_I-closed subset of X (τ, I), then $A\cap Y$ is a β_{I_Y}-closed set in Y (τ_Y, I_Y).*

Proof: If A is β_I-closed, then A^C is β_I-open. By Theorem 1.11, $A^C\cap Y$ is β_{I_Y}-open. Hence, $A\cap Y = \left(A^C\cap Y\right)^C$ is β_{I_Y}-closed in Y. □

The next remark is clear. We shall be using it in showing some of the succeeding theorems.

Remark 1.5. *Let (X,τ,I) be an ideal topological space and $Y\subseteq X$. Then $I_Y = \{A\cap Y : A\in I\}$ is a subset of I.*

Proof: If A is a β_{I_Y}-open set in Y, then there exists an open set $O\in\tau_Y$ with $O - A\in I_Y$ and $A - \text{cl}(\text{int}(\text{cl}(O)))\in I_Y$. Because $\tau_Y\subseteq\tau$ and by Remark 1.5, there exists an open set $O\in\tau$ with $O - A\in I$ and $A - \text{cl}(\text{int}(\text{cl}(O)))\in I$. Thus, A is β_I-open in $X(\tau, I)$.

The converse follows from Theorem 1.11. □

The next statement, Lemma 1.8, characterizes β_I-open sets in subspaces.

Lemma 1.8. *Let (X,τ,I) be an ideal τ_ζ-space, Y $(\subseteq X)$ be clopen, and $\tau_Y\subseteq\tau$. If $A\subseteq Y$, then A is β_{I_Y}-open in $Y(\tau_Y, I_Y)$ if and only if it is β_I-open in $X(\tau, I)$.*

Proof: If A is a β_{I_Y}-open set in Y, then there exists an open set $O\in\tau_Y$ with $O - A\in I_Y$ and $A - \text{cl}(\text{int}(\text{cl}(O)))\in I_Y$. Because $\tau_Y\subseteq\tau$ and by Remark 1.5, there exists an open set $O\in\tau$ with $O - A\in I$ and $A - \text{cl}(\text{int}(\text{cl}(O)))\in I$. Thus, A is β_I-open in $X(\tau, I)$.

The converse follows from Theorem 1.11. □

The next statement, Theorem 1.12, provides a way of determining the closure of a set in the subspace.

Theorem 1.12. *Let (X,τ,I) be an ideal τ_ζ-space, Y be clopen, and $\tau_Y\subseteq\tau$. If $A\subseteq X$, then $\text{cl}_{\beta_{I_Y}}(A\cap Y) = \text{cl}_{\beta_I}(A)\cap Y$.*

Proof: Since $\text{cl}_{\beta_I}(A)$ is a β_I-closed set in X, by Lemma 1.8 $\text{cl}_{\beta_I}(A)\cap Y$ is a β_{I_Y}-closed set in Y. Hence, $\text{cl}_{\beta_{I_Y}}(A\cap Y) \supseteq \text{cl}_{\beta_{I_Y}}\left(\text{cl}_{\beta_I}(A)\cap Y\right) = \text{cl}_{\beta_I}(A)\cap Y$. But, $\text{cl}_{\beta_{I_Y}}(A\cap Y) = \text{cl}_{\beta_I}(A\cap Y)\subseteq \text{cl}_{\beta_I}(A)\cap\text{cl}_{\beta_I}(Y) = \text{cl}_{\beta_I}(A)\cap Y$. Therefore, $\text{cl}_{\beta_{I_Y}}(A\cap Y) = \text{cl}_{\beta_I}(A)\cap Y$. □

Definition 1.14. *Let (X,τ,I) be an ideal topological space, and (Y,τ_Y,I_Y) be a subspace. A pair of subsets, say A and B, of X is said to be β_{I_Y}-separated in Y if $cl_{\beta_{I_Y}}(A)\cap B = \varnothing = A\cap cl_{\beta_Y}(B)$, where $cl_{\beta_Y}(B) = cl_\beta(B)\cap Y$.*

Definition 1.15. *Let (X,τ,I) be an ideal topological space, and (Y,τ_Y,I_Y) be a subspace. A subset A of Y is said to be β_{I_Y}-connected if it cannot be expressed as a union of two β_{I_Y}-separated sets. The subspace Y is said to be β_{I_Y}-connected if it is β_{I_Y}-connected as a subset.*

The next statement, Theorem 1.13, says that if two sets are separated in the mother space, then they are also separated in the subspace.

Theorem 1.13. *Let (X,τ) be a τ_ζ-space, I be an ideal, Y $(\subseteq X)$ be clopen, and $\tau_Y\subseteq\tau$. If A and B are β_I-separated in X, then they are β_{I_Y}-separated in Y.*

Proof: If A and B are β_I-separated in X, then by Theorem 1.12 $\varnothing = \text{cl}_{\beta_I}(A)\cap B = \left(\text{cl}_{\beta_I}(A)\cap B\right)\cap Y = \left(\text{cl}_{\beta_I}(A)\cap Y\right)\cap B = \text{cl}_{\beta_{I_Y}}(A)\cap B$ and $\varnothing = A\cap\text{cl}_\beta(B) = \left(A\cap\text{cl}_\beta(B)\right)\cap Y = A\cap\left(\text{cl}_\beta(B)\cap Y\right) = A\cap\text{cl}_{\beta_Y}(B)$. Thus, A and B are β_{I_Y}-separated. □

The next statement, Remark 1.6, says that if two non-empty sets, which expresses X as a disjoint union, is β_I-separated, then one must be β-open and the other must be β_I-open.

Remark 1.6. *Let (X,τ) be a topological space and I be an ideal. If X is β_I-separated (say, $X = A\cup B$ with $A\neq\varnothing$, $B\neq\varnothing$, and $cl_{\beta_I}(A)\cap B = \varnothing = A\cap cl_\beta(B)$), then A is β-open while B is β_I-open.*

To see this, if A and B is β_I-separated, then $\text{cl}_{\beta_I}(A)\cap B = \varnothing$ and $A\cap\text{cl}_\beta(B) = \varnothing$. Hence, $A^C = \text{cl}_\beta(B)$ and $B^C = \text{cl}_{\beta_I}(A)$. Thus, A^C is β-closed and B^C is β_I-closed. Accordingly, A is β-open and B is β_I-open.

The next statement, Theorem 1.14, characterizes β_I-connected spaces.

Theorem 1.14. *Let (X, τ) be a topological space and I be an ideal X. Then, X is β_I-connected if and only if it cannot be expressed as a union of two a non-empty disjoint sets in which one is a β-open set and the other is a β_I-open set.*

Proof: Suppose that X is β_I-connected, and we can express X as a union of two non-empty disjoint β-open set and β_I-open set, say $A \cup B = X$ (with A, a β-open set, and B, a β_I-open set) and $A \cap B = \emptyset$. If $A \cup B = X$ and $A \cap B = \emptyset$, then $A^C = B$ and $B^C = A$. Since A is β-open, B is β-closed. Also, since B is β_I-open, A is β_I-closed. Hence, $\text{cl}_{\beta_I}(A) \cap B = A \cap B = \emptyset$ and $A \cap \text{cl}_\beta(B) = A \cap B = \emptyset$. Thus, the pair A and B is β_I-separated. This is a contradiction.

The converse follows from Remark 1.6. □

The next statement, Theorem 1.15, says that two separated set cannot contain portions of a connected set.

Theorem 1.15. *Let (X, τ) be a topological space, I be an ideal X, and A be a β_I-connected set. If $A \subseteq H \cup G$ where H and G is a pair of β_I-separated sets, then either $A \subseteq H$ or $A \subseteq G$.*

Proof: Suppose that to the contrary, $A = (A \cap H) \cup (A \cap G)$ with $A \cap H \neq \emptyset$ and $A \cap G \neq \emptyset$. Since H and G is a pair of β_I-separated sets, $\text{cl}_{\beta_I}(A \cap H) \cap (A \cap G) \subseteq \text{cl}_{\beta_I}(H) \cap (G) = \emptyset$ and $(A \cap H) \cap \text{cl}_\beta(A \cap G) \subseteq H \cap \text{cl}_\beta(G) = \emptyset$. Thus, $\text{cl}_{\beta_I}(A \cap H) \cap (A \cap G) = \emptyset$ and $(A \cap H) \cap \text{cl}_\beta(A \cap G) = \emptyset$. Therefore, A can be expressed as a union of two β_I-separated sets $A \cap H$ and $A \cap G$. A contradiction. □

The next statement, Theorem 1.16, says that subsets of each of two separated sets are also separated.

Theorem 1.16. *Let (X, τ) be a topological space, I be an ideal in X, and, A and B be β_I-separated sets. If $C \subseteq A$ $(C \neq \emptyset)$ and $D \subseteq B$ $(D \neq \emptyset)$, then C and D are also β_I-separated.*

Proof: Suppose that A and B are β_I-separated. Then $\text{cl}_{\beta_I}(A) \cap B = \emptyset$ and $A \cap \text{cl}_\beta(B) = \emptyset$. Thus, $\text{cl}_{\beta_I}(C) \cap D \subseteq \text{cl}_{\beta_I}(A) \cap B = \emptyset$ and $C \cap \text{cl}_\beta(D) = A \cap \text{cl}_\beta(B) = \emptyset$. Hence, $\text{cl}_{\beta_I}(C) \cap D = \emptyset = C \cap \text{cl}_\beta(D)$. Therefore, C and D is β_I-separated. □

5. Conclusion

With the important concepts and results which intertwined with those introduced by other authors, this chapter is very interesting. The construction of the different theorems were realized using the definitions or properties of β-open sets, β_I-compact spaces, β_I^*-hyperconnected spaces, β_I-separated spaces. Also, some properties focusing on generalizing ideals in ideal topological space theory were realized.

Author details

Glaisa T. Catalan[1†], Michael P. Baldado Jr[2*†] and Roberto N. Padua[3†✠]

1 Negros Oriental State University—Siaton Campus, Siaton, Negros Oriental, Philippines

2 Mathematics Department, Negros Oriental State University—Main Campus, Dumaguete City, Philippines

3 Commission on Higher Education (CHED), Philippines

*Address all correspondence to: michael.baldadojr@norsu.edu.ph

† These authors contributed equally.

✠ Deceased.

References

[1] Levine N. Semi-open sets and semi-continuity in topological spaces. The American Mathematical Monthly. 1963; **70**(1):36-41

[2] Njástad O. On some classes of nearly open sets. Pacific Journal of Mathematics. 1965;**15**(3):961-970

[3] Mashhour AS, Abd El-Monsef ME, El-Deeh SN. On pre-continuous and weak pre-continuous mappings. Proceedings of the Mathematical and Physical Society of Egypt. 1982;**53**:47-53

[4] Stone MH. Applications of the theory of boolean rings to general topology. Transactions of the American Mathematical Society. 1937;**41**(3):375-481

[5] Mubarki AM, Al-Rshudi MM, Al-Juhani MA. β^*-open sets and β^*-continuity in topological spaces. Journal of Taibah University for Science. 2014; **8**(2):142-148

[6] Kannan K, Nagaveni N. On $\hat{\beta}$-generalized closed sets in topological spaces. International Journal of Mathematical Analysis. 2012;**6**(57): 2819-2828

[7] Abd El-Monsef ME. β-open sets and β-continuous mappings. Bulletin of the Faculty of Science. Assiut University. 1983;**12**:77-90

[8] Akdag M, Ozkan A. On soft β-open sets and soft β-continuous functions. The Scientific World Journal. Hindawi. 2014;**2014**:1-6. DOI: 10.1155/2014/843456

[9] Arockiarani I, Arokia LA. Generalized soft $g\beta$-closed sets and soft $gs\beta$-closed sets in soft topological spaces. International Journal of Mathematical Archive. 2013;**4**(2):1-7

[10] Dugundji J. Topology. Boston: Ally and Bacon; 1966

[11] Kuratowski K. Topology. Poland: Scientific Publishers. 1996

[12] Morris SA. Topology without Tears. Australia: University of New England; 1989

[13] Tahiliani S. Operation approach to β-open sets and applications. Mathematical Communications. 2011; **16**(2):577-591

[14] Bhattacharyya P. Semi-generalized closed sets in topology. Indian Journal of Mathematics. 1987;**29**(3):375-382

[15] Skowron A. On topology information systems. Bulletin of the Polish Academy of Sciences. 1989;**3**: 87-90

[16] Abid MY. Non semi-pre-denseness in topological spaces. Journal of Kerbala University. 2007;**5**(2):159-163

[17] El-Mabhouh A, Mizyed A. On the topology generated by βc-open sets. International Journal of Mathematical Sciences and Engineering Applications (IJMSEA). 2015;**9**(1):223-232

[18] Kuratowski K. Topologie. Bulletin of the American Mathematical Society. 1934;**40**:787-788

[19] Vaidyanathaswamy R. Set Topology, Chelsea, New York. Lubbock, Texas: University of New Mexico, Albuquerque, New Mexico Texas Technological College; 1960

[20] Shravan K, Tripathy BC. Generalised closed sets in multiset topological space. Proyecciones (Antofagasta). 2018;**37**(2):223-237

[21] Shravan K, Tripathy BC. Multiset ideal topological spaces and local functions. Proyecciones (Antofagasta). 2018;**37**(4):699-711

[22] Chandra Tripathy B, Acharjee S. On (γ, δ)-bitopological semi-closed set via topological ideal. Proyecciones (Antofagasta). 2014;**33**(3):245-257

[23] Tripathy BC, Ray GC. Mixed fuzzy ideal topological spaces. Applied Mathematics and Computation. 2013; **220**:602-607

[24] Michael FI. On semi-open sets with respect to an ideal. European Journal of Pure and Applied Mathematics. 2013; **6**(1):53-58

[25] Newcomb RL. Topologies which are compact modulo an ideal [Ph.d. dissertation]. Santa Barbara, CA: University of California; 1967

[26] Erdal E, Takashi N. *-hyperconnected ideal topological spaces. Analele ştiinţifice Ale Universităţii "Al.I. Cuza" Din Iaçsi (S.N.) Matematică. 2012;**LVIII**:121-129

[27] Maheshwari SN, Thakur SS. On α-compact spaces. Bulletin of the Institute of Mathematics. 1985;**13**(4):341-347

[28] Nasef AA, Radwan AE, Esmaeel RB. Some properties of α-open sets with respect to an ideal. International Journal of Pure and Applied Mathematics. 2015; **102**(4):613-630

[29] Abd El-Monsef ME, Nasef AA, Radwan AE, Ibrahem FA, Esmaeel RB. Some properties of semi-open sets with respect to an ideal. International Electronic Journal of Pure and Applied Mathematics. 2015;**9**(3):167-179. DOI: 10.12732/iejpam.v9i3.6

Section 6

Riemannian Submersions

Chapter 8

Clairaut Submersion

Sanjay Kumar Singh and Punam Gupta

Abstract

In this chapter, we give the detailed study about the Clairaut submersion. The fundamental notations are given. Clairaut submersion is one of the most interesting topics in differential geometry. Depending on the condition on distribution of submersion, we have different classes of submersion such as anti-invariant, semi-invariant submersions etc. We describe the geometric properties of Clairaut anti-invariant submersions and Clairaut semi-invariant submersions whose total space is a Kähler, nearly Kähler manifold. We give condition for Clairaut anti-invariant submersion to be a totally geodesic map and also study Clairaut anti-invariant submersions with totally umbilical fibers. We also give the conditions for the semi-invariant submersions to be Clairaut map and also for Clairaut semi-invariant submersion to be a totally geodesic map. We also give some illustrative example of Clairaut anti-invariant and semi-invariant submersion.

Keywords: Riemannian submersion, nearly Kähler manifolds, Kähler manifolds, anti-invariant submersion, semi-invariant submersion, clairaut submersion, totally geodesic maps

1. Introduction

Riemannian submersion between two Riemannian manifolds was first introduced by O'Neill [1] and Gray [2]. After that Watson [3] introduced almost Hermitian submersions. Later, the notion of anti-invariant submersions and Lagrangian submersion from almost Hermitian manifolds onto Riemannian manifolds were introduced by Sahin [4] and studied by Taştan [5, 6], Gündüzalp [7], Beri et al. [8], Ali and Fatima [9], in which the fibers of submersion are anti-invariant with respect to the almost complex structure of total manifold. After that several new types of Riemannian submersions were defined and studied such as semi-invariant submersion [10, 11], slant submersion [12, 13], generic submersion [14–17], hemi-slant submersion [18], semi-slant submersion [19], pointwise slant submersion [20–22] and conformal semi-slant submersion [23]. Also, these kinds of submersions were considered in different kinds of structures such as nearly Kähler, Kähler, almost product, para-contact, Sasakian, Kenmotsu, cosymplectic and etc. In book [24], we find the recent developments in this field.

In 1735, A.C. Clairaut [25] obtained the very important result in the theory of surfaces, which is Clairaut's theorem and stated that for any geodesic α on a surface of revolution S, the function $r \sin\theta$ is constant along α, where r is the distant from a point on the surface to the rotation axis and θ is the angle between α and the meridian through α. Bishop [26] introduced the idea of Riemannian submersions and gave a necessary and sufficient conditions for a Riemannian submersion to be Clairaut. Allison [27] considered Clairaut semi-Riemannian submersions and showed that such submersions have interesting applications in the static space-times.

In [28], Tastan and Gerdan gave new Clairaut conditions for anti-invariant submersions whose total manifolds are Sasakian and Kenmotsu and got many interesting results. In [29], Tastan and Aydin studied Clairaut anti-invariant submersions whose total manifolds are cosymplectic. Gündüzalp [30] introduced Clairaut anti-invariant submersions from a paracosymplectic manifold and gave characterization theorems. In [31], Lee et al. studied Clairaut anti-invariant submersions whose total manifolds are Kähler.

Kähler manifolds [32, 33] have an especially rich geometric and topological structure because of Kähler identity. Kähler manifolds are very important in differential geometry, which has applications in several different fields such as supersymmetric gauge theory and superstring theory in theoretical physics, signal processing in information geometry. The simplest example of Kähler manifold is a complex Euclidean space $\mathbb{C}^n$ with the standard Hermitian metric.

Nearly Kähler manifolds introduced by Gray and Hervella [32], are the geometrically interesting class among the sixteen classes of almost Hermitian manifolds. The geometrical meaning of nearly Kähler condition is that the geodesics on the manifolds are holomorphically planar curves. Gray [2] studied nearly Kähler manifolds broadly and gave example of a non-Kählerian nearly Kähler manifold, which is 6-dimensional sphere.

Motivated by this, the authors [34] studied Clairaut anti-invariant submersions from nearly Kähler manifolds onto Riemannian manifolds with some examples and obtained conditions for Clairaut Riemannian submersion to be totally geodesic map. The authors investigated conditions for the Clairaut anti-invariant submersions to be a totally umbilical map. The authors [34] studied Clairaut semi-invariant submersions from Kähler manifolds onto Riemannian manifolds with some examples. The authors also obtained conditions for Clairaut semi-invariant Riemannian submersion to be totally geodesic map and investigated conditions for the semi-invariant submersion to be a Clairaut map.

2. Almost complex manifold

An almost complex structure on a smooth manifold M is a smooth tensor field φ of type $(1,1)$ such that $\varphi^2 = -I$. A smooth manifold equipped with such an almost complex structure is called an almost complex manifold. An almost complex manifold (M, φ) endowed with a chosen Riemannian metric g satisfying

$$g(\varphi X, \varphi Y) = g(X, Y) \tag{1}$$

for all $X, Y \in TM$, is called an almost Hermitian manifold.

An almost Hermitian manifold M is called a nearly Kähler manifold [2] if

$$(\nabla_X \varphi)Y + (\nabla_Y \varphi)X = 0 \tag{2}$$

for all $X, Y \in TM$. If $(\nabla_X \varphi)Y = 0$ for all $X, Y \in TM$, then M is known as Kähler manifold [33]. Every Kähler manifold is nearly Kähler but converse need not be true.

3. Riemannian submersion

Definition 1.1 [1, 35] Let (M, g_m) and (N, g_n) be Riemannian manifolds, where $\dim(M) = m$, $\dim(N) = n$ and $m > n$. A Riemannian submersion $\pi : M \to N$ is a map of M onto N satisfying the following axioms:

i. π has maximal rank.

ii. The differential π_* preserves the lengths of horizontal vectors.

For each $q \in N$, $\pi^{-1}(q)$ is an $(m-n)$-dimensional Riemannian submanifold of M. The submanifolds $\pi^{-1}(q)$, $q \in N$, are called fibers. A vector field on M is called vertical if it is always tangent to fibers. A vector field on M is called horizontal if it is always orthogonal to fibers. A vector field X on M is called basic if X is horizontal and π-related to a vector field X' on N, that is, $\pi_* X_p = X'_{\pi_*(p)}$ for all $p \in M$. We denote the projection morphisms on the distributions $\ker\pi_*$ and $(\ker\pi_*)^\perp$ by $\mathcal{V}$ and $\mathcal{H}$, respectively. The sections of $\mathcal{V}$ and $\mathcal{H}$ are called the vertical vector fields and horizontal vector fields, respectively. So

$$\mathcal{V}_p = T_p(\pi^{-1}(q)), \qquad \mathcal{H}_p = T_p(\pi^{-1}(q))^\perp.$$

The second fundamental tensors of all fibers $\pi^{-1}(q)$, $q \in N$ gives rise to tensor field T and A in M defined by O'Neill [1] for arbitrary vector field E and F, which is

$$T_E F = \mathcal{H}\nabla^M_{\mathcal{V}E}\mathcal{V}F + \mathcal{V}\nabla^M_{\mathcal{V}E}\mathcal{H}F, \tag{3}$$

$$A_E F = \mathcal{H}\nabla^M_{\mathcal{H}E}\mathcal{V}F + \mathcal{V}\nabla^M_{\mathcal{H}E}\mathcal{H}F, \tag{4}$$

where $\mathcal{V}$ and $\mathcal{H}$ are the vertical and horizontal projections.

To discuss geodesics, we need a linear connection. We denote the Levi-Civita connection on M by $\hat{\nabla}$ and the adapted connection of the submersion by ∇. From Eqs. (3) and (4), we have

$$\nabla_V W = T_V W + \hat{\nabla}_V W, \tag{5}$$

$$\nabla_V X = \mathcal{H}\nabla_V X + T_V X, \tag{6}$$

$$\nabla_X V = A_X V + \mathcal{V}\nabla_X V, \tag{7}$$

$$\nabla_X Y = \mathcal{H}\nabla_X Y + A_X Y, \tag{8}$$

for all $V, W \in \Gamma(\ker\pi_*)$ and $X, Y \in \Gamma(\ker\pi_*)^\perp$, where $\mathcal{V}\nabla_V W = \hat{\nabla}_V W$. If X is basic, then $A_X V = \mathcal{H}\nabla_V X$.

It is easily seen that for $p \in M$, $U \in \mathcal{V}_p$ and $X \in \mathcal{H}_p$ the linear operators

$$T_U, A_X : T_pM \to T_pM$$

are skew-symmetric, that is,

$$g(A_X E, F) = -g(E, A_X F) \text{ and } g(T_U E, F) = -g(E, T_U F), \tag{9}$$

for all $E, F \in T_pM$. We also see that the restriction of T to the vertical distribution $T|_{\ker\pi_* \times \ker\pi_*}$ is exactly the second fundamental form of the fibers of π. Since T_U is skew-symmetric, therefore π has totally geodesic fibers if and only if $T \equiv 0$.

Let $\pi : (M, g_m) \to (N, g_n)$ be a smooth map between Riemannian manifolds. Then the differential π_* of π can be observed as a section of the bundle $Hom(TM, \pi^{-1}TN) \to M$, where $\pi^{-1}TN$ is the bundle which has fibers $(\pi^{-1}TN)_x = T_{f(x)}N$. $Hom(TM, \pi^{-1}TN)$ has a connection ∇ induced from the Riemannian connection ∇^M and the pullback connection ∇^N [36, 37]. Then the second fundamental form of π is given by

$$(\nabla\pi_*)(E,F) = \nabla^N_E \pi_* F - \pi_*\left(\nabla^M_E F\right), \quad \text{for all} E, F \in \Gamma(TM). \tag{10}$$

We also know that π is said to be totally geodesic map [36] if $(\nabla\pi_*)(E,F) = 0$, for all $E, F \in \Gamma(TM)$.

4. Clairaut submersion from Riemannian manifold

Let S be a revolution surface in $\mathbb{R}^3$ with rotation axis L. For any $p \in S$, we denote by $r(p)$ the distance from p to L. Given a geodesic $\alpha : J \subset \mathbb{R} \to S$ on S, let $\theta(t)$ be the angle between $\alpha(t)$ and the meridian curve through $\alpha(t), t \in I$. A well-known Clairaut's theorem [25] named after Alexis Claude de Clairaut, says that for any geodesic on S, the product $r \sin\theta$ is constant along α, i.e., it is independent of t. For proof, see [38, p.183]. In the theory of Riemannian submersions, Bishop [26] introduced the notion of Clairaut submersion in the following way:

Definition 1.2 [26] A Riemannian submersion $\pi : (M,g) \to (N,g_n)$ is called a Clairaut submersion if there exists a positive function r on M, which is known as the girth of the submersion, such that, for any geodesic α on M, the function $(r \circ \alpha)\sin\theta$ is constant, where, for any t, $\theta(t)$ is the angle between $\dot{\alpha}(t)$ and the horizontal space at $\alpha(t)$.

For further use, we are stating one important result of Bishop.

Theorem 1.1 [26] A curve h in M is a geodesic if and only if $\dot{X} + 2A_X U + T_U U = 0$ and $\nabla_E U + T_U X = 0$, where $\dot{h}(t) = E = X + U$, X is horizontal and U is vertical.

Bishop also gave the following necessary and sufficient condition for a Riemannian submersion to be a Clairaut submersion, which is

Theorem 1.2 [26] Let $\pi : (M,g) \to (N,g_n)$ be a Riemannian submersion with connected fibers. Then, π is a Clairaut submersion with $r = e^f$ if and only if each fiber is totally umbilical and has the mean curvature vector field $H = -gradf$, where $gradf$ is the gradient of the function f with respect to g.

Proof: Let $\pi : M \to N$ be a Riemannian submersion. For a geodesic h in M, we use $\dot{h}(s) = E = X + U$, where X is horizontal and U is vertical. and $\ell = \left\|\dot{h}(s)\right\|^2$. Let $\theta(s)$ be the angle between $\dot{h}(s)$ and the horizontal space at $h(s)$. Then

$$g(X(s), X(s)) = \ell \cos^2\theta(s), \tag{11}$$

$$g(U(s), U(s)) = \ell \sin^2\theta(s). \tag{12}$$

Differentiating (12), we get

$$g\left(\nabla_{\dot{h}(s)} U(s), U(s)\right) = \ell \sin\theta(s)\cos\theta(s)\frac{d\theta(s)}{ds}. \tag{13}$$

Using Theorem 1.1, (13) becomes

$$-g\left(T_{U(s)} X(s), U(s)\right) = \ell \sin\theta(s)\cos\theta(s)\frac{d\theta(s)}{ds}. \tag{14}$$

Since T_U is skew-symmetric, so form above equation, we have

$$g\left(T_{U(s)} U(s), X(s)\right) = \ell \sin\theta(s)\cos\theta(s)\frac{d\theta(s)}{ds}. \tag{15}$$

Now, π is a Clairaut submersion with $r = e^f$ if and only if $\frac{d}{ds}\left(e^{f \circ h} sin\theta\right) = 0$. Using (12, 15) in $\frac{d}{ds}\left(e^{f \circ h} sin\theta\right) = 0$, we have

$$g(U(s), U(s))\frac{d}{ds}(f \circ h)(s) + g\left(T_{U(s)}U(s), X(s)\right) = 0, \tag{16}$$

$$g(U(s), U(s))g(gradf, E(s)) + g\left(T_{U(s)}U(s), X(s)\right) = 0. \tag{17}$$

Consider any geodesic h on M with initial vertical tangent vector, so $gradf$ turns out to be horizontal. Therefore, the function f is constant on any fiber, the fibers being connected. Therefore (17) reduces to

$$g(U(s), U(s))g(gradf, X(s)) + g\left(T_{U(s)}U(s), X(s)\right) = 0, \tag{18}$$

$$g(U(s), U(s))gradf + T_{U(s)}U(s) = 0. \tag{19}$$

Setting $U = U_1 + U_2$, where U_1, U_2 are vertical vector fields and using the fact that T is symmetric for vertical vector fields, we obtain

$$g(U_1, U_2)gradf + T_{U_1}U_2 = 0 \tag{20}$$

holds for all vertical vector fields U_1, U_2..

Since the restriction of T to the vertical distribution $T|_{\ker\pi_* \times \ker\pi_*}$ is exactly the second fundamental form of the fibers of π. It means that any fiber is totally umbilical with mean curvature vector field $H = -gradf$.

Conversely, suppose the fibers are totally umbilic with normal curvature vector field $H = -gradf$ so that we have

$$g(U, U)H + T_U U = 0. \tag{21}$$

Since $gradf$ is orthogonal to fibers, so

$$g(U, U)g(gradf, E) = -g(U, U)g(H, X) = -g(T_U U, X). \tag{22}$$

Since (18) holds. so $(r \circ h)sin\theta$ is constant along any geodesic h.

Example 1.1 [24] Consider the warped product manifold $M_1 \times_f M_2$ of Riemannian manifolds (M_1, g_1) and (M_2, g_2), where $f : M_1 \to (0, \infty)$. The fibers of the first projection $p_1 : M_1 \times_f M_2 \to M_1$ are totally umbilical with mean curvature vector field $H = -grad\left(logf^{1/2}\right)$. Thus, if M_2 is connected, p_1 is a Clairaut submersion with $r = f^{1/2}$.

5. Anti-invariant Riemannian submersion

Definition 1.3 [39] Let (M, φ, g) be an almost Hermitian manifold and N be a Riemannian manifold with Riemannian metric g_n. Suppose that there exists a Riemannian submersion $\pi : M \to N$, such that the vertical distribution $\ker\pi_*$ is anti-invariant with respect to φ, i.e., $\varphi\ker\pi_* \subseteq \ker\pi_*^{\perp}$. Then, the Riemannian submersion π is called an anti-invariant Riemannian submersion. We will briefly call such submersions as anti-invariant submersions.

Let π be an anti-invariant Riemannian submersion from nearly Kähler manifold (M, φ, g_m) onto Riemannian manifold (N, g_n). For any arbitrary tangent vector fields U and V on M, we set

$$(\nabla_U \varphi)V = P_U V + Q_U V \tag{23}$$

where $P_U V, Q_U V$ denote the horizontal and vertical part of $(\nabla_U \varphi)V$, respectively. Clearly, if M is a Kähler manifold then $P = Q = 0$.

If M is a nearly Kähler manifold then P and Q satisfy

$$P_U V = -P_V U, \qquad Q_U V = -Q_V U. \tag{24}$$

Consider

$$(\ker \pi_*)^\perp = \varphi \ker \pi_* \oplus \mu,$$

where μ is the complementary distribution to $\varphi \ker \pi_*$ in $(ker\pi_*)^\perp$ and $\varphi\mu \subset \mu$. For $X \in \Gamma(ker\pi_*)^\perp$, we have

$$\varphi X = \alpha X + \beta X, \tag{25}$$

where $\alpha X \in \Gamma(\ker \pi_*)$ and $\beta X \in \Gamma(\mu)$. If $\mu = 0$, then an anti-invariant submersion is known as Lagrangian submersion.

5.1 Anti-invariant Clairaut submersions from nearly Kähler manifolds

In this section, we give new Clairaut conditions for anti-invariant submersions from nearly Kähler manifolds after giving some auxiliary results.

Theorem 1.3 [34] Let π be an anti-invariant submersion from a nearly Kähler manifold (M, φ, g) onto a Riemannian manifold (N, g_n). If $h : J \subset \mathbb{R} \to M$ is a regular curve and $U(s)$ and $X(s)$ are the vertical and horizontal parts of the tangent vector field $\dot{h}(s) = W$ of $h(s)$, respectively, then h is a geodesic if and only if along h

$$A_X \varphi U + A_X \beta X + T_U \beta X + \mathcal{V}\nabla_X \alpha X + T_U \varphi U + \hat{\nabla}_U \alpha X = 0, \tag{26}$$

$$\mathcal{H}\left(\nabla_{\dot{h}} \varphi U + \nabla_{\dot{h}} \beta X\right) + A_X \alpha X + T_U \alpha X = 0. \tag{27}$$

Proof: Let π be an anti-invariant submersion from a nearly Kähler manifold (M, φ, g) onto a Riemannian manifold (N, g_n). Since $\varphi^2 \dot{h} = -\dot{h}$. Taking the covariant derivative of this and using (2), we have

$$(\nabla_{\dot{h}} \varphi)\varphi \dot{h} + \varphi\left(\nabla_{\dot{h}} \varphi \dot{h}\right) = -\nabla_{\dot{h}} \dot{h}. \tag{28}$$

Since $U(s)$ and $X(s)$ are the vertical and horizontal parts of the tangent vector field $\dot{h}(s) = W$ of $h(s)$, that is, $\dot{h} = U + X$. So (28) becomes

$$\begin{aligned} -\nabla_{\dot{h}} \dot{h} &= \varphi(\nabla_{U+X} \varphi(U + X)) + P_{\dot{h}} \varphi \dot{h} + Q_{\dot{h}} \varphi \dot{h} \\ &= \varphi(\nabla_U \varphi U + \nabla_X \varphi U + \nabla_U \varphi X + \nabla_X \varphi X) + P_{\dot{h}} \varphi \dot{h} + Q_{\dot{h}} \varphi \dot{h} \\ &= \varphi(\nabla_U \varphi U + \nabla_X \varphi U + \nabla_U(\alpha X + \beta X) + \nabla_X(\alpha X + \beta X)) \\ &\quad + P_{\dot{h}} \varphi \dot{h} + Q_{\dot{h}} \varphi \dot{h}. \end{aligned} \tag{29}$$

Using (5)–(8) in (29), we get

$$-\nabla_{\dot{h}}\dot{h} = \varphi\big(\mathcal{H}(\nabla_{\dot{h}}\varphi U + \nabla_{\dot{h}}\beta X) + A_X\alpha X + A_X\beta X + A_X\varphi U \\ +T_U\beta X + T_U\alpha X + \mathcal{V}\nabla_X\alpha X + T_U\varphi U + \hat{\nabla}_U\alpha X\big) + P_{\dot{h}}\varphi\dot{h} + Q_{\dot{h}}\varphi\dot{h}. \tag{30}$$

Let $Y, Z \in TM$. Since $\varphi^2 Z = -Z$, on differentiation, we have

$$\varphi(\nabla_Y\varphi Z) + (\nabla_Y\varphi)\varphi Z = -\nabla_Y Z,$$
$$\varphi^2(\nabla_Y Z) + \varphi(\nabla_Y\varphi)Z + (\nabla_Y\varphi)\varphi Z = -\nabla_Y Z,$$

using (23) in above, we obtain

$$\varphi(P_Y Z + Q_Y Z) = -P_Y\varphi Z - Q_Y\varphi Z. \tag{31}$$

By (31), we have

$$\varphi\left(P_{\dot{h}}\varphi\dot{h} + Q_{\dot{h}}\varphi\dot{h}\right) = P_{\dot{h}}\dot{h} + Q_{\dot{h}}\dot{h},$$

since P and Q are skew-symmetric, so

$$\varphi\left(P_{\dot{h}}\varphi\dot{h} + Q_{\dot{h}}\varphi\dot{h}\right) = 0. \tag{32}$$

Using (32) and equating the vertical and horizontal part of (30), we obtain

$$\mathcal{V}\varphi\nabla_{\dot{h}}\dot{h} = A_X\varphi U + A_X\beta X + T_U\beta X + \mathcal{V}\nabla_X\alpha X + T_U\varphi U + \hat{\nabla}_U\alpha X,$$
$$\mathcal{H}\varphi\nabla_{\dot{h}}\dot{h} = \mathcal{H}(\nabla_{\dot{h}}\varphi U + \nabla_{\dot{h}}\beta X) + A_X\alpha X + T_U\alpha X.$$

By using above equations, we can say that h is geodesic if and only if (26, 27) hold.

Theorem 1.4 [34] Let π be an anti-invariant submersion from a nearly Kähler manifold (M, φ, g) onto a Riemannian manifold (N, g_n). Also, let $h : J \subset \mathbb{R} \to M$ be a regular curve and $U(s)$ and $X(s)$ are the vertical and horizontal parts of the tangent vector field $\dot{h}(s) = W$ of $h(s)$. Then π is a Clairaut submersion with $r = e^f$ if and only if along h

$$g(gradf, X)g(U, U) = g\left(\mathcal{H}\nabla_{\dot{h}}\beta X + A_X\alpha X + T_U\alpha X + P_{\dot{h}(s)}U, \varphi U\right).$$

Proof: Let $h : J \subset \mathbb{R} \to M$ be a geodesic on M and $\ell = \left\|\dot{h}(s)\right\|^2$. Let $\theta(s)$ be the angle between $\dot{h}(s)$ and the horizontal space at $h(s)$. Then

$$g(X(s), X(s)) = \ell\cos^2\theta(s), \tag{33}$$
$$g(U(s), U(s)) = \ell\sin^2\theta(s). \tag{34}$$

Differentiating (34), we get

$$2g\left(\nabla_{\dot{h}(s)}U(s), U(s)\right) = 2\ell\sin\theta(s)\cos\theta(s)\frac{d\theta(s)}{ds}. \tag{35}$$

Using (1) in (35), we have

$$g\left(\mathcal{H}\nabla_{\dot{h}(s)}\varphi U(s), \varphi U(s)\right) - g\left(\left(\nabla_{\dot{h}(s)}\varphi\right)U(s), \varphi U(s)\right) = \ell\sin\theta(s)\cos\theta(s)\frac{d\theta(s)}{ds}.$$

Now by use of (23), we have

$$g\left(\mathcal{H}\nabla_{\dot{h}(s)}\varphi U(s), \varphi U(s)\right) - g\left(P_{\dot{h}(s)}U + Q_{\dot{h}(s)}U, \varphi U(s)\right) = \ell \sin\theta(s)\cos\theta(s)\frac{d\theta(s)}{ds}.$$

Along the curve h, using Theorem 1.3, we obtain

$$-g\left(\mathcal{H}\nabla_{\dot{h}}\beta X + A_X\alpha X + T_U\alpha X + P_{\dot{h}(s)}U, \varphi U(s)\right) = \ell \sin\theta(s)\cos\theta(s)\frac{d\theta(s)}{ds}.$$

Now, π is a Clairaut submersion with $r = e^f$ if and only if $\frac{d}{ds}\left(e^f \sin\theta\right) = 0$. Therefore

$$e^f\left(\frac{df}{ds}\sin\theta + \cos\theta\frac{d\theta}{ds}\right) = 0,$$

$$e^f\left(\frac{df}{ds}\ell\sin^2\theta + \ell\sin\theta\cos\theta\frac{d\theta}{ds}\right) = 0.$$

So, we obtain

$$\frac{df}{ds}(h(s))g(U(s), U(s)) = g\left(\mathcal{H}\nabla_{\dot{h}}\beta X + A_X\alpha X + T_U\alpha X + P_{\dot{h}(s)}U, \varphi U(s)\right). \tag{36}$$

Since $\frac{df}{ds}(h(s)) = g\left(gradf, \dot{h}(s)\right) = g(gradf, X)$. Therefore by using (36), we get the result.

Theorem 1.5 [34] Let π be an Clairaut anti-invariant submersion from a nearly Kähler manifold (M, φ, g) onto a Riemannian manifold (N, g_n) with $r = e^f$. Then

$$A_{\varphi W}\varphi X + Q_W\varphi X = X(f)W$$

for $X \in (\ker\pi_*)^{\perp}$, $W \in \ker\pi_*$ and φW is basic.

Proof: Let π be an anti-invariant submersion from a nearly Kähler manifold (M, φ, g) onto a Riemannian manifold (N, g_n) with $r = e^f$. We know that any fiber of Riemannian submersion π is totally umbilical if and only if

$$T_V W = g(V, W)H, \tag{37}$$

for all $V, W \in \Gamma(\ker\pi_*)$, where H denotes the mean curvature vector field of any fiber in M. By using Theorem 1.2 and (37), we have

$$T_V W = -g(V, W)gradf. \tag{38}$$

Let $X \in \mu$ and $V, W \in \Gamma(\ker\pi_*)$, then by using (1) and (2), we have

$$g(\nabla_V\varphi W, \varphi X) = g(\varphi\nabla_V W + (\nabla_V\varphi)W, \varphi X) = g(\nabla_V W, X) + g(P_V W + Q_V W, \varphi X). \tag{39}$$

By using (1), we have

$$g(\varphi Y, Z) = -g(Y, \varphi Z),$$

where $Y, Z \in TM$. Taking covariant derivative of above, we get

DOI: http://dx.doi.org/10.5772/intechopen.101427

$$g((\nabla_X\varphi)Y,Z) = -g(Y,(\nabla_X\varphi)Z),$$

using (23), we get

$$\begin{aligned} g(P_XY+Q_XY,Z) &= -g(Y,P_XZ+Q_XZ) \\ &= g(Y,P_ZX+Q_ZX). \end{aligned} \tag{40}$$

Using (40), we have

$$g(P_W\varphi X+Q_W\varphi X,V) = g(\varphi X,P_VW+Q_VW). \tag{41}$$

Using (5), (38), (41) in (39), we have

$$g(\nabla_V\varphi W,\varphi X) = -g(V,W)(gradf,X)+g(V,Q_W\varphi X).$$

Since φW is basic, so $\mathcal{H}\nabla_V\varphi W = A_{\varphi W}V$, therefore we have

$$g(A_{\varphi W}V,\varphi X) = -g(V,W)(gradf,X)+g(V,Q_W\varphi X),$$
$$g(V,A_{\varphi W}\varphi X)+g(V,Q_W\varphi X) = g(V,W)(gradf,X) \tag{42}$$

because A is skew-symmetric. By using (42), we get the result.

Theorem 1.6 [34] Let π be a Clairaut anti-invariant submersion from a nearly Kähler manifold (M,φ,g) onto a Riemannian manifold (N,g_n) with $r=e^f$ and $gradf \in \varphi\ker\pi_*$. Then either f is constant on $\varphi\ker\pi_*$ or the fibers of π are 1-dimensional.

Proof: Using (5) and (38), we have

$$g(\nabla_VW,\varphi U) = -g(V,W)g(gradf,\varphi U),$$

where $U,V,W\in\Gamma(\ker\pi_*)$. Since $g(W,\varphi U)=0$. therefore we have

$$g(W,\nabla_V\varphi U) = g(V,W)g(gradf,\varphi U). \tag{43}$$

By use of (1) and (23) in (43), we get

$$g(W,Q_VU)-g(\varphi W,\nabla_VU) = g(V,W)g(gradf,\varphi U).$$

By using (5), we obtain

$$g(W,Q_VU)-g(\varphi W,T_VU) = g(V,W)g(gradf,\varphi U).$$

Now, using (38), we get

$$g(W,Q_VU)+g(V,U)g(gradf,\varphi W) = g(V,W)g(gradf,\varphi U) \tag{44}$$

Take $V=U$ in (44), we have

$$g(V,V)g(gradf,\varphi W) = g(V,W)g(gradf,\varphi V). \tag{45}$$

Interchange V with W in (45), we have

$$g(W,W)g(gradf,\varphi V) = g(V,W)g(gradf,\varphi W). \tag{46}$$

By (45) and (46), we have

$$g^2(V,W)g(gradf,\varphi V)=g(V,V)g(W,W)g(gradf,\varphi V).$$

Therefore either f is constant on $\varphi \ker\pi_*$ or $V=aW$, where a is constant (by using Schwarz's Inequality for equality case).

Corollary 1.1 [34] Let π be a Clairaut anti-invariant submersion from a nearly Kähler manifold (M,φ,g) onto a Riemannian manifold (N,g_n) with $r=e^f$ and $gradf \in \varphi\ker\pi_*$. If $\dim(\ker\pi_*)>1$, then the fibers of π are totally geodesic if and only if $A_{\varphi W}\varphi X+Q_W\varphi X=0$ for $W\in\ker\pi_*$ such that φW is basic and $X\in\mu$.

Proof: By Theorem 1.5 and Theorem 1.6, we get the result.

Corollary 1.2 [34] Let π be an Clairaut Lagrangian submersion from a nearly Kähler manifold (M,φ,g) onto a Riemannian manifold (N,g_n) with $r=e^f$. Then either the fibers of π are 1-dimensional or they are totally geodesic.

Proof: Let π be an Clairaut Lagrangian submersion from a Kähler manifold (M,φ,g) onto a Riemannian manifold (N,g_n) with $r=e^f$. Then $\mu=\{0\}$, so $A_{\varphi W}\varphi X+Q_W\varphi X=0$ always.

Now, we discuss some examples for Clairaut anti-invariant submersions from a nearly Kähler manifold.

Example 1.2 [34] Let $(\mathbb{R}^4,\varphi,g)$ be a nearly Kähler manifold endowed with Euclidean metric g on $\mathbb{R}^4$ given by

$$g=\sum_{i=1}^{4}dx_i^2$$

and canonical complex structure

$$\varphi(x_j)=\begin{cases}-x_{j+1} & j=1,3\\ x_{j-1} & j=2,4\end{cases}.$$

The φ-basis is $\left\{e_i=\frac{\partial}{\partial x_i}|i=1,2,3,4\right\}$. Let $(\mathbb{R}^3,g_1)$ be a Riemannian manifold endowed with metric $g_1=\sum_{i=1}^{3}dy_i^2$.

i. Consider a map $\pi:(\mathbb{R}^4,\varphi,g)\to(\mathbb{R}^3,g_1)$ defined by

$$\pi(x_1,x_2,x_3,x_4)=\left(\frac{x_1+x_2}{\sqrt{2}},x_3,x_4\right).$$

Then by direct calculations, we have

$$\ker\pi_*=span\left\{X_1=\left(\frac{\partial}{\partial x_1}-\frac{\partial}{\partial x_2}\right)\right\},$$

$$(\ker\pi_*)^{\perp}=span\left\{X_2=\left(\frac{\partial}{\partial x_1}+\frac{\partial}{\partial x_2}\right),X_3=\frac{\partial}{\partial x_3},X_4=\frac{\partial}{\partial x_4}\right\}$$

and $\varphi X_1=-X_2$, therefore $\varphi(\ker\pi_*)\subset(\ker\pi_*)^{\perp}$. Thus, we can say that π is an anti-invariant Riemannian submersion. Since the fibers of π are 1-dimensional, therefore fibers are totally umbilical.

Consider the Koszul formula for Levi-Civita connection ∇ for $\mathbb{R}^4$

$$2g(\nabla_X Y, Z) = Xg(Y, Z) + Yg(Z, X) - Zg(X, Y) - g([Y, Z], X) - g([X, Z], Y) + g([X, Y], Z)$$

for all $X, Y, Z \in \mathbb{R}^4$. By simple calculations, we obtain

$$\nabla_{e_i} e_j = 0 \quad \text{for all}\, i, j = 1, 2, 3, 4.$$

Hence $T_X Y = T_Y X = T_X X = 0$ for all $X, Y \in \Gamma(\ker \pi_*)$. Therefore fibers of π are totally geodesic. Thus π is Clairaut trivially.

ii. Consider a map $\pi : (\mathbb{R}^4, \varphi, g) \to (\mathbb{R}^3, g_1)$ defined by

$$\pi(x_1, x_2, x_3, x_4) = \left(\sqrt{x_1^2 + x_2^2}, x_3, x_4\right).$$

Then by direct calculations, we have

$$\ker \pi_* = span\left\{X_1 = \left(\frac{x_2}{\sqrt{x_1^2 + x_2^2}}\frac{\partial}{\partial x_1} - \frac{x_1}{\sqrt{x_1^2 + x_2^2}}\frac{\partial}{\partial x_2}\right)\right\},$$

$$(\ker \pi_*)^\perp = span\left\{X_2 = \left(\frac{x_2}{\sqrt{x_1^2 + x_2^2}}\frac{\partial}{\partial x_1} + \frac{x_1}{\sqrt{x_1^2 + x_2^2}}\frac{\partial}{\partial x_2}\right), X_3 = \frac{\partial}{\partial x_3}, X_4 = \frac{\partial}{\partial x_4}\right\}$$

and $\varphi X_1 = -X_2$, therefore $\varphi(\ker \pi_*) \subset (\ker \pi_*)^\perp$. Thus, we can say that π is an anti-invariant Riemannian submersion. Since the fibers of π are 1-dimensional, therefore fibers are totally umbilical. By using Koszul formula, we obtain

$$\nabla_{e_i} e_j = 0 \quad \text{for all}\, i, j = 1, 2, 3, 4.$$

Hence

$$T_{X_1} X_1 = -\left(\frac{x_2}{\sqrt{x_1^2 + x_2^2}}\frac{\partial}{\partial x_1} + \frac{x_1}{\sqrt{x_1^2 + x_2^2}}\frac{\partial}{\partial x_2}\right).$$

Now, for the function $f = \ln\left(\sqrt{x_1^2 + x_2^2}\right)$ on $(\mathbb{R}^4, \varphi, g)$, the gradient of f with respect to g is given by

$$gradf = \sum_{i,j=1}^{4} g^{ij}\frac{\partial f}{\partial x_i}\frac{\partial}{\partial x_j} = \frac{x_2}{\sqrt{x_1^2 + x_2^2}}\frac{\partial}{\partial x_1} + \frac{x_1}{\sqrt{x_1^2 + x_2^2}}\frac{\partial}{\partial x_2}.$$

Therefore for $X_1 \in \Gamma(\ker \pi_*)$, $T_{X_1} X_1 = -gradf$. Since $\|X_1\| = 1$, so $T_{X_1} X_1 = -\|X_1\|^2 gradf$. By using Theorem 1.2, we can say that π is an proper Clairaut anti-invariant submersion with $r = e^f$ for $f = \ln\left(\sqrt{x_1^2 + x_2^2}\right)$.

Remark: From all results of this section, we can easily find conditions for anti-invariant Clairaut Submersions from Kähler manifolds.

6. Semi-invariant Riemannian submersion

Definition 1.4 Let (M, φ, g) be an almost Hermitian manifold and N be a Riemannian manifold with Riemannian metric g_n. A Riemannian submersion

$\pi : M \to N$ is called a semi-invariant Riemannian submersion [11] if there is a distribution $\mathcal{D}_1 \subseteq \ker\pi_*$ such that

$$\ker\pi_* = \mathcal{D}_1 \oplus \mathcal{D}_2 \text{ and } \varphi(\mathcal{D}_1) = \mathcal{D}_1, \quad \varphi(\mathcal{D}_2) \subseteq (\ker\pi_*)^{\perp},$$

where $\mathcal{D}_2$ is orthogonal complementary to $\mathcal{D}_1$ in $\ker\pi_*$. For $V \in \Gamma(ker\pi_*)$, we have

$$\varphi V = \phi V + \omega V, \tag{47}$$

where $\phi V \in \Gamma(\mathcal{D}_1)$ and $\omega V \in \Gamma(\varphi\mathcal{D}_2)$.

Definition 1.5 A semi-invariant Riemannian submersion π is said to be a Lagrangian Riemannian submersion [4] if $\varphi(\ker\pi_*) = (\ker\pi_*)^{\perp}$. Hence, if π is a Lagrangian Riemannian submersion then for any $V \in \Gamma(ker\pi_*)$, $\varphi V = \omega V$, $\phi V = 0$ and for $X \in \Gamma(ker\pi_*)^{\perp}$, $\varphi X = \alpha X$, $\beta X = 0$.

6.1 Semi-invariant Clairaut submersions from Kähler manifolds

In this section, we give new Clairaut conditions for semi-invariant submersions from Kähler manifolds after giving some auxiliary results.

Theorem 1.7 [40] Let π be a semi-invariant submersion from a Kähler manifold (M, φ, g) onto a Riemannian manifold (N, g_n). If $h : J \subset \mathbb{R} \to M$ is a regular curve and $U(s)$ and $X(s)$ are the vertical and horizontal parts of the tangent vector field $\dot{h}(s) = W$ of $h(s)$, respectively, then h is a geodesic if and only if along h

$$\mathcal{V}\nabla_X\phi U + \mathcal{V}\nabla_X\alpha X + A_X\omega U + A_X\beta X + \hat{\nabla}_U\phi U + T_U\beta X + T_U\omega U + \hat{\nabla}_U\alpha X = 0, \tag{48}$$

$$A_X\phi U + A_X\alpha X + \mathcal{H}(\nabla_{\dot{h}}\omega U + \nabla_{\dot{h}}\beta X) + T_U\phi U + T_U\alpha X = 0. \tag{49}$$

Proof: Let π be a semi-invariant submersion from a Kähler manifold (M, φ, g) onto a Riemannian manifold (N, g_n). Since $\varphi^2\dot{h} = -\dot{h}$. Taking the covariant derivative of this and using (2), we have

$$\varphi\left(\nabla_{\dot{h}}\varphi\dot{h}\right) = -\nabla_{\dot{h}}\dot{h}. \tag{50}$$

Since $U(s)$ and $X(s)$ are the vertical and horizontal parts of the tangent vector field $\dot{h}(s) = W$ of $h(s)$, that is, $\dot{h} = U + X$. So (50) becomes

$$\begin{aligned} -\nabla_{\dot{h}}\dot{h} &= \varphi(\nabla_{U+X}\varphi(U+X)) \\ &= \varphi(\nabla_U\varphi U + \nabla_X\varphi U + \nabla_U\varphi X + \nabla_X\varphi X) \\ &= \varphi(\nabla_U(\phi U + \omega U) + \nabla_X(\phi U + \omega U) + \nabla_U(\alpha X + \beta X) + \nabla_X(\alpha X + \beta X)). \end{aligned} \tag{51}$$

Using (5)–(8) in (51), we get

$$\begin{aligned} -\nabla_{\dot{h}}\dot{h} = \varphi(&\mathcal{H}(\nabla_{\dot{h}}\varphi U + \nabla_{\dot{h}}\beta X) + A_X\alpha X + A_X\beta X + A_X\varphi U \\ &+ T_U\beta X + T_U\alpha X + \mathcal{V}\nabla_X\alpha X + T_U\varphi U + \hat{\nabla}_U\alpha X). \end{aligned} \tag{52}$$

Equating the vertical and horizontal part of (52), we obtain

$$\mathcal{V}\varphi\nabla_{\dot{h}}\dot{h} = \mathcal{V}\nabla_X\phi U + \mathcal{V}\nabla_X\alpha X + A_X\omega U + A_X\beta X + \hat{\nabla}_U\phi U + T_U\beta X + T_U\omega U + \hat{\nabla}_U\alpha X,$$

$$\mathcal{H}\varphi\nabla_{\dot{h}}\dot{h} = A_X\phi U + A_X\alpha X + \mathcal{H}(\nabla_{\dot{h}}\omega U + \nabla_{\dot{h}}\beta X) + T_U\phi U + T_U\alpha X.$$

By using above equations, we can say that h is geodesic if and only if (48) and (49) hold.

Theorem 1.8 [40] Let π be a semi-invariant submersion from a Kähler manifold (M,φ,g) onto a Riemannian manifold (N,g_n). Also, let $h : J\subset\mathbb{R}\to M$ be a regular curve. $U(s)$ and $X(s)$ are the vertical and horizontal parts of the tangent vector field $\dot{h}(s) = W$ of $h(s)$. Then π is a Clairaut submersion with $r = e^f$ if and only if along h

$$g(gradf,X)g(U,U) = g(\mathcal{H}\nabla_{\dot{h}}\beta X + A_X\alpha X + T_U\alpha X, \omega U) + g(\mathcal{V}\nabla_X\alpha X + A_X\beta X + T_U\beta X + \hat{\nabla}_U\alpha X, \phi U).$$

Proof: Let $h : J\subset\mathbb{R}\to M$ be a geodesic on M and $\ell = \left\|\dot{h}(s)\right\|^2$. Let $\theta(s)$ be the angle between $\dot{h}(s)$ and the horizontal space at $h(s)$. Then

$$g(X(s),X(s)) = \ell\cos^2\theta(s), \tag{53}$$

$$g(U(s),U(s)) = \ell\sin^2\theta(s). \tag{54}$$

Differentiating (54), we get

$$2g\left(\nabla_{\dot{h}(s)}U(s),U(s)\right) = 2\ell\sin\theta(s)\cos\theta(s)\frac{d\theta(s)}{ds}. \tag{55}$$

Using (1) in (55), we have

$$g\left(\nabla_{\dot{h}(s)}\varphi U(s),\varphi U(s)\right)) = \ell\sin\theta(s)\cos\theta(s)\frac{d\theta(s)}{ds}.$$

Now by use of (47), we have

$$g\left(\nabla_{\dot{h}(s)}\phi U(s),\varphi U(s)\right) + g\left(\nabla_{\dot{h}(s)}\omega U(s),\varphi U(s)\right) = \ell\sin\theta(s)\cos\theta(s)\frac{d\theta(s)}{ds}$$

Along the curve h, using Theorem 1.7 and (5)–(8), we obtain

$$\ell\sin\theta(s)\cos\theta(s)\frac{d\theta(s)}{ds} = -g(\mathcal{H}\nabla_{\dot{h}}\beta X + A_X\alpha X + T_U\alpha X, \omega U)(s) - g(\mathcal{V}\nabla_X\alpha X + A_X\beta X + T_U\beta X + \hat{\nabla}_U\alpha X, \phi U)(s)$$

Now, π is a Clairaut submersion with $r = e^f$ if and only if $\frac{d}{ds}\left(e^f\sin\theta\right) = 0$. Therefore

$$e^f\left(\frac{df}{ds}\sin\theta + \cos\theta\frac{d\theta}{ds}\right) = 0,$$

$$e^f\left(\frac{df}{ds}\ell\sin^2\theta + \ell\sin\theta\cos\theta\frac{d\theta}{ds}\right) = 0.$$

So, we obtain

$$\frac{df}{ds}(h(s))g(U(s),U(s)) = g\left(\mathcal{H}\nabla_{\dot{h}}\beta X + A_X\alpha X + T_U\alpha X, \omega U\right)(s) + g\left(\mathcal{V}\nabla_X\alpha X + A_X\beta X + T_U\beta X + \hat{\nabla}_U\alpha X, \phi U\right)(s), \tag{56}$$

Since $\frac{df}{ds}(h(s)) = \dot{h}[g](s) = g\left(gradf, \dot{h}(s)\right) = g(gradf, X)$. Therefore by using (56), we get the result.

Theorem 1.9 [40] Let π be a Clairaut semi-invariant submersion from a Kähler manifold (M, φ, g) onto a Riemannian manifold (N, g_n) with $r = e^f$. Then

$$g(A_{\omega W}V, \beta X) + g(\mathcal{H}\nabla_V\phi W, \beta X) + g(\mathcal{V}\nabla_V\omega W, \alpha X) + g\left(\hat{\nabla}_V\phi W, \alpha X\right) = -g(V, W)(Xf)$$

for $X \in \Gamma\mu$, $V, W \in \Gamma(\mathcal{D}_2)$ and ωW is basic.

Proof: Let π be a Clairaut semi-invariant submersion from a Kähler manifold (M, φ, g) onto a Riemannian manifold (N, g_n) with $r = e^f$. We know that any fiber of Riemannian submersion π is totally umbilical if and only if

$$T_V W = g(V, W)H, \tag{57}$$

for all $V, W \in \Gamma(\ker\pi_*)$, where H denotes the mean curvature vector field of any fiber in M. By using Theorem 1.2 and (57), we have

$$T_V W = -g(V, W)gradf. \tag{58}$$

Let $X \in \mu$ and $V, W \in \Gamma(\ker\pi_*)$, then by using (1) and (2), we have

$$g(\nabla_V\varphi W, \varphi X) = g(\varphi\nabla_V W + (\nabla_V\varphi)W, \varphi X) = g(\nabla_V W, X). \tag{59}$$

Using (5), (58) in (59), we have

$$g(\nabla_V\varphi W, \varphi X) = -g(V, W)(gradf, X).$$

Since ωW is basic, so $\mathcal{H}\nabla_V\omega W = A_{\omega W}V$, therefore we have

$$g(A_{\omega W}V, \beta X) + g(\mathcal{H}\nabla_V\phi W, \beta X) + g(\mathcal{V}\nabla_V\omega W, \alpha X) + g\left(\hat{\nabla}_V\phi W, \alpha X\right) = -g(V, W)(gradf, X). \tag{60}$$

Theorem 1.10 [40] Let π be a Clairaut semi-invariant submersion from a Kähler manifold (M, φ, g) onto a Riemannian manifold (N, g_n) with $r = e^f$ and $V, W \in \Gamma(\mathcal{D}_1)$ Then $gradf \in \Gamma(\varphi\ker\pi_*)$.

Proof: Let $V, W \in \Gamma(\mathcal{D}_1)$ and $X \in \Gamma(\mu)$. Using (5), (47) and (58) in

$$\nabla_V\varphi U = \varphi\nabla_V U + (\nabla_V\varphi)U,$$

we have

$$T_V\phi U + \mathcal{V}\nabla_V\phi U = \alpha T_V U + \beta T_V U + \phi\mathcal{V}\nabla_V U + \omega\mathcal{V}\nabla_V U,$$

which gives

$$-g(V, \phi U)g(gradf, X) = g(V, U)g(gradf, \varphi X). \tag{61}$$

By interchanging U and V in (61) and adding the resulting equation with (61), we get

$$g(V,U)g(gradf,\varphi X)=0,$$

which gives $g(gradf,\varphi X)=0$. Therefore $gradf\in\Gamma(\varphi\ker\pi_*)$.

Theorem 1.11 [40] Let π be a Clairaut semi-invariant submersion from a Kähler manifold (M,φ,g) onto a Riemannian manifold (N,g_n) with $r=e^f$ and $gradf\in\Gamma(\varphi\ker\pi_*)$. Then either f is constant on $\varphi\ker\pi_*$ or the fibers of π are 1-dimensional.

Proof: Let $U,V\in\Gamma(\mathcal{D}_2)$. Using (5) and (58), we have

$$g(\nabla_V U,\varphi U)=-g(V,U)g(gradf,\varphi U),$$

which gives

$$g(\varphi\nabla_V U,U)=g(V,U)g(gradf,\varphi U),$$

since $\ker\pi_*$ is integrable, so we have

$$g(\varphi\nabla_U V,U)=g(V,U)g(gradf,\varphi U),$$

which equals to

$$g(\nabla_U\varphi V-(\nabla_U\varphi)V,U)=g(V,U)g(gradf,\varphi U).$$

Since $g(\varphi V,U)=0$. therefore we have

$$g(\varphi V,\nabla_U U)=-g(V,U)g(gradf,\varphi U). \tag{62}$$

By using (5) in (62), we obtain

$$g(\varphi V,T_U U)=-g(V,U)g(gradf,\varphi U).$$

Now, using (58), we get

$$g(U,U)g(gradf,\varphi V)=g(V,U)g(gradf,\varphi U). \tag{63}$$

Interchanging V and U in (63), we have.

$$g(V,V)g(gradf,\varphi U)=g(V,U)g(gradf,\varphi V). \tag{64}$$

By (63) and (64), we have

$$g^2(V,U)g(gradf,\varphi U)=g(V,V)g(U,U)g(gradf,\varphi U).$$

Therefore either f is constant on $\varphi\ker\pi_*$ or $V=aU$, where a is constant (by using Schwarz's Inequality for equality case).

Since $\ker\pi_*$ is CR-submanifold of Kähler manifold (M,φ,g), therefore by using [41], Theorem 6.1, p. 96], we can state that.

Theorem 1.12 [40] Let π be a Clairaut semi-invariant submersion from a Kähler manifold (M,φ,g) onto a Riemannian manifold (N,g_n) with $r=e^f$. If $\dim\mathcal{D}_2>1$, then fibers are totally geodesic.

Corollary 1.3 [40] Let π be a Clairaut Lagrangian submersion from a Kähler manifold (M, φ, g) onto a Riemannian manifold (N, g_n) with $r = e^f$. Then either the fibers of π are 1-dimensional or they are totally geodesic.

Lastly, we discuss some examples for Clairaut semi-invariant submersions [40] from a Kähler manifold.

Example 1.3 Every Clairaut anti-invariant submersion from a Kähler manifold onto a Riemannian manifold is a Clairaut semi-invariant submersion with $\mathcal{D}_1 = \{0\}$.

Example 1.4 Let $(\mathbb{R}^6, \varphi, g)$ be a Kähler manifold endowed with Euclidean metric g on $\mathbb{R}^6$ given by

$$g = \sum_{i=1}^{6} dx_i^2$$

and canonical complex structure

$$\varphi(x_j) = \begin{cases} -x_{j+1} & j = 1, 3, 5 \\ x_{j-1} & j = 2, 4, 6 \end{cases}.$$

The φ-basis is $\left\{e_i = \frac{\partial}{\partial x_i} | i = 1, \ldots, 6\right\}$. Let $(\mathbb{R}^3, g_1)$ be a Riemannian manifold endowed with metric $g_1 = \sum_{i=1}^{3} dy_i^2$.

Consider a map $\pi : (\mathbb{R}^6, \varphi, g) \rightarrow (\mathbb{R}^3, g_1)$ defined by

$$\pi(x_1, x_2, x_3, x_4, x_5, x_6) = \left(\frac{x_1 + x_2}{\sqrt{2}}, x_3, x_4\right).$$

Then by direct calculations, we have

$$\ker\pi_* = span\left\{X_1 = \left(\frac{\partial}{\partial x_1} - \frac{\partial}{\partial x_2}\right), X_2 = \frac{\partial}{\partial x_5}, X_3 = \frac{\partial}{\partial x_6}\right\},$$

$$(\ker\pi_*)^\perp = span\left\{V_1 = \left(\frac{\partial}{\partial x_1} + \frac{\partial}{\partial x_2}\right), V_2 = \frac{\partial}{\partial x_3}, V_3 = \frac{\partial}{\partial x_4}\right\}$$

and $\varphi X_1 = -V_1$, $\varphi X_2 = -X_3$, $\varphi X_3 = -X_2$ therefore $\mathcal{D}_1 = span\{X_2, X_3\}$ and $\mathcal{D}_2 = span\{X_1\}$. Thus, we can say that π is a semi-invariant Riemannian submersion.

Consider the Koszul formula for Levi-Civita connection ∇ for $\mathbb{R}^6$

$$2g(\nabla_X Y, Z) = Xg(Y, Z) + Yg(Z, X) - Zg(X, Y) - g([Y, Z], X) - g([X, Z], Y) + g([X, Y], Z)$$

for all $X, Y, Z \in \mathbb{R}^6$. By simple calculations, we obtain

$$\nabla_{e_i} e_j = 0 \quad \text{for all } i, j = 1, \ldots, 6.$$

Hence $T_X Y = T_Y X = T_X X = 0$ for all $X, Y \in \Gamma(\ker\pi_*)$. Therefore fibers of π are totally geodesic. Thus π is Clairaut trivially.

Classification

2010 mathematics subject classification. 53C12, 53C15, 53C20, 53C55

Author details

Sanjay Kumar Singh[1]* and Punam Gupta[2]

1 Indian Institute of Science Education and Research, Bhopal, India

2 Dr. Harisingh Gour University, Sagar, India

*Address all correspondence to: sanjayks@iiserb.ac.in

References

[1] O'Neill B. The fundamental equations of a submersion. Michigan Mathematical Journal. 1966;**13**:458-469

[2] Gray A. Nearly Kähler manifolds. Journal of Differential Geometry. 1970; **6**:283-309

[3] Watson B. Almost Hermitian submersions. Journal of Differential Geometry. 1976;**11**:147-165

[4] Sahin B. Anti-invariant Riemannian submersion from almost Hermitian manifolds. Central Europian Journal of Mathematics. 2010;**8**(3):437-447

[5] Tastan HM. On Lagrangian submersions. Hacettepe Journal of Mathematics and Statistics. 2014;**43**(6): 993-1000

[6] Tastan HM, Özdemir F, Sayar C. On anti-invariant Riemannian submersions whose total manifolds are locally product Riemannian. Journal of Geometry. 2017;**108**:411-422

[7] Gündüzalp Y. Anti-invariant Riemannian submersions from almost product Riemannian manifolds. Mathematical Sciences and Applications E-Notes. 2013;**1**:58-66

[8] Beri A, Kupeli EI, Murathan C. Anti-invariant Riemannian submersions from Kenmotsu manifolds onto Riemannian manifolds. Turkish Journal of Mathematics. 2016;**40**(3): 540-552

[9] Ali S, Fatima T. Anti-invariant Riemannian submersions from nearly Kähler manifolds. Univerzitet u Nišu. 2013;**27**:1219-1235

[10] Özdemir F, Sayar C, Tastan HM. Semi-invariant submersions whose total manifolds are locally product Riemannian. Quaestiones Mathematicae. 2017;**40**(7):909-926

[11] Sahin B. Semi-invariant submersion from almost Hermitian manifolds. Canadian Mathematical Bulletin. 2011; **56**:173-182

[12] Küpeli Erken I, Murathan C. On slant submersions for cosymplectic manifolds. Bulletin of the Korean Mathematical Society. 2014;**51**(6):1749-1771

[13] Sahin B. Slant submersions from almost Hermitian manifolds. Bulletin mathématique de la Société des Sciences Mathématiques de Roumanie. 2011;**54 (102)**(1):93-105

[14] Akyol MA. Generic Riemannian submersions from almost product Riemannian manifolds. GUJ Sci. 2017; **30**(3):89-100

[15] Ali S, Fatima T. Generic Riemannian submersions. Tamkang Journal of Mathematics. 2013;**44**(4):395-409

[16] Chen BY. Differential geometry of real submanifolds in Kähler manifold. Monatshefte für Mathematik. 1981;**91**: 257-274

[17] Ronsse GB. Generic and skew CR-submanifolds of a Kähler manifold. Bulletin of the Institute of Mathematics, Academia Sinica. 1990;**18**:127-141

[18] Tastan HM, Sahin B, Yanan S. Hemi-slant submersions. Mediterranean Journal of Mathematics. 2016;**13**(4):2171-2184

[19] Park KS, Prasad R. Semi-slant submersions. Bulletin of the Korean Mathematical Society. 2013;**50**(3):951-962

[20] Aykurt SA, Ergüt M. Pointwise slant submersions from cosymplectic manifolds. Turkish Journal of Mathematics. 2016;**40**(3):582-593

[21] Chen BY, Garay O. Pointwise slant submanifolds in almost Hermitian

manifolds. Turkish Journal of Mathematics. 2012;**364**:630-640

[22] Lee JW, Sahin B. Pointwise slant submersions. Bulletin of the Korean Mathematical Society. 2014;**51**:1115-1126

[23] Akyol MA. Conformal semi-slant submersions. International Journal of Geometric Methods in Modern Physics. 2017;**14**(7):1750114

[24] Sahin B. Riemannian Submersions, Riemannian Maps in Hermitian Geometry and their Applications. Cambridge: Elsevier Academic Press; 2017

[25] Clairaut AC. Mémoires de l'Académie royale des sciences. Mem. Acad. Paris Pour. 1735;**1733**:86

[26] Bishop RL. Clairaut submersions, Differential Geometry (in Honor of Kentaro Yano). Tokyo: Kinokuniya; 1972. pp. 21-31

[27] Allison D. Lorentzian Clairaut submersions. Geometriae Dedicata. 1996;**63**(3):309-319

[28] Tastan HM, Gerdan S. Clairaut anti-invariant submersions from Sasakian and Kenmotsu manifolds. Mediterranean Journal of Mathematics. 2017;**14**(6):235. 17 pp

[29] Tastan HM, Aydin SG. Clairaut anti-invariant submersions from cosymplectic manifolds. Honam Mathematical Journal. 2019;**41**(4): 707-724

[30] Gündüzalp Y. Anti-invariant pseudo-Riemannian submersions and Clairaut submersions from paracosymplectic manifolds. Mediterranean Journal of Mathematics. 2019;**16**(4):94. 18 pp

[31] Lee J, Park JH, Shahin B, Song DY. Einstein conditions for the base of anti-invariant Riemannian submersions and Clairaut submersions. Taiwanse Journal of Mathematics. 2015;**19**(4): 1145-1160

[32] Gray A, Hervella M. The sixteen classes of almost Hermitian manifolds and their linear invariants. Annali di matematica pura ed applicata. 1980;**123**: 35-58

[33] Yano K. Structures on manifolds, Series in Pure Mathematics, 3. Singapore: World Scientific; 1984

[34] Gupta P, Rai AK. Clairaut anti-invariant submersion from nearly Kähler manifold. arXiv. Mat. Vesnik, Accepted. 2020. 2010.05552

[35] O'Neill B. Semi-Riemannian Geometry with Application to Relativity. New York: Academic Press; 1983

[36] Baird P, Wood JC. Harmonic Morphisms Between Riemannian Manifolds, London Mathematical Society Monographs. Vol. 29. Oxford: Oxford University Press, The Clarendon Press; 2003

[37] Bejancu A. Geometry of CR-Submanifolds. Dordrecht: Kluwer Academic Publishers; 1986

[38] Pressley A. Elementary Differential Geometry. Springer-Verlag London: Springer Undergraduate Mathematics Series; 2001

[39] Lee JW. Anti-invariant $\xi^{\perp}$-Riemannian submersions from almost contact manifolds. Hacettepe Journal of Mathematics and Statistics. 2013;**42**(3): 231-241

[40] Gupta P, Singh SK. Clairaut semi-invariant submersion from Kähler manifold. Afrika Mat. Accepted. 2020. DOI: 10.13140/RG.2.2.30240.30725

[41] Yano K, Kon M. CR-Submanifolds of Kählerian and Sasakian Manifolds. Progress in Mathematics, 30. Birkhäuser, Boston: Mass; 1983